Introduction To Business Statistics Through R Software

Software

Editor IJSMI

Published by IJSMI, 2023.

INTRODUCTION TO BUSINESS STATISTICS THROUGH R SOFTWARE

First edition. July 2, 2023.

ISBN: 979-8223254966

Written by Editor IJSMI.

Introduction to Business Statistics through R software

Preface

Statistical methods are now widely used in different fields such as Business and Management, Economics, Biological, Physical sciences and including the new fields such as Data Science and Machine Learning. The data which form the basis for the statistical methods helps us to take scientific and informed decisions. Statistical methods deal with the collection, compilation, analysis and making inference from the data.

This book deals with the statistical methods which are useful in Business and Management decision making. The methods include Probability, Sampling, Correlation, Regression and Hypothesis Testing, Time Series, Forecasting and Non-Parametric tests and advanced statistical models. The book uses open source R statistical software to carry out different statistical analysis with sample datasets.

This book is third in series of Statistics books by the Author. Some of the contents are adopted from the author's previous statistical book introduction to statistical methods and non-parametric methods.

Editor

International Journal of Statistics and Medical Informatics

www.ijsmi.com/book.php[1]

1. http://www.ijsmi.com/book.php

Chapter 1 - Introduction

(Some of the contents are adopted from Author's previous book – Introduction to Statistical Methods)

Statistics is defined as collection, compilation, analysis and interpretation of data to draw meaningful conclusions and to make informed & scientific decisions.

Statistical methods help us to answers the commons questions like the following

1. How much percentage of males contributes the Population of United States
2. What is daily average sales figure of online site - xyz.com
3. What is defects rate of items manufactured today
4. What will be price of Apple stock after one month
5. What is the chance that Brazil football team winning the football world cup
6. What is the size of male foot ware which needs to be produced more in USA

The following chapters of the books help the readers to answer the questions like the ones mentioned above with the use of statistical methods, tools and techniques.

1.1 Basic Concepts

Data

Data is the most basic form of information and it is present in the form of numbers, text, image and sound.

Variables

Variables are quantities whose values vary according to defined rule or scope. For example, age of a person can be defined by the term or variable "Age" and it can take the numeric values ranging from say 0 to 120 or gender of a patient can be represented by the term or variable "gender" and it can take text values such as "Male" or "Female" in a dataset.

Constants

Constants are quantities which can take a constant value in a dataset. Example - Normal body temperature

Population

Population is a collection of all possible data related to a specific parameter or point of interest. The population can be definite in numbers or countable or indefinite in number or non-countable. Examples of population are student population of New York, online customer population, or population of stars in the Universe.

Sample

A subset drawn from the given population is called as sample. Examples of sample - sample of 100 online customers drawn from the online customer population or a student sample of size 100 drawn from the student population of New York.

MEASUREMENT SCALES

Data can be measured in the nominal, ordinal, interval and ratio scales

Nominal Scale

Nominal scales measurement is used when the data provides information about characteristic or name of an entity (variable) which is being measured. Example of nominal scale measurements are name of the patient, place of birth etc. It cannot be useful for further mathematical treatment (addition, subtraction etc.) as it lacks relative or ratio property.

Ordinal scale

Ordinal Measurement scale is used when the data indicates only the order or rank and it cannot be useful for further mathematical treatment as it lacks relative or ratio property. Difference between rank 7 and rank 8 is not same as difference between 1 and 2. Example of ordinal measurement is ranking of customer preference on product or service offered.

Interval scale

Interval measurement scale is used when the data is used to represent the relative property between the values not the ratio property of the data in consideration. Ratings of customer satisfaction can be measured in interval scale. Difference between rating 2 and 3 on a scale is same as difference between 4 and 5 but at the same time it can't represent that the satisfaction of the customer with 4 rating is as twice as customer with a rating 2.

Ratio scale

Ratio scale represents the ratio property of the data being measured and it is useful for further mathematical treatment (addition, subtraction etc.). Number of orders received (150) from young customers is more than 2 times the middle age group customers (75)

Discrete variable

If the variable of a dataset contains only integer value or measured as integer value then the variable is called as discrete variable

Continuous variable

If the variable of a dataset contains real numbers then the variable is called as continuous variable

Distribution

Distribution of a dataset provides information about the spread of data in the dataset. Distribution of variable is defined as discrete if the variable under consideration is of discrete type and same way it is termed as continuous if the variable under consideration is of continuous type. Distribution of a dataset is represented through frequency tables and graphical form (Histogram etc.).

Frequency Table

Frequency table provides us the information on how many times each data value is present in the dataset in case of discrete data and In case of the continuous data it will provide us information between two ranges on measurement scale. In both cases, the frequency table provides information about all the values which are present in the data set.

Gender	US Population (March 2020)	Frequency %
Male	164,596,841	49.4%
Female	168,815,541	50.6%
Total	333,412,382	100%

Table1.1 - Frequency table for gender

Histogram

Histogram is a graphical display of the how the values of a dataset are distributed in the dataset. It displays the frequency with respect to each of the data point in case of discrete data or interval in case of continuous data. The horizontal axis includes the entire range of values of the data set i.e. it starts with the minimum value and ends with the maximum value and vertical axis represents the frequency in each interval or data point

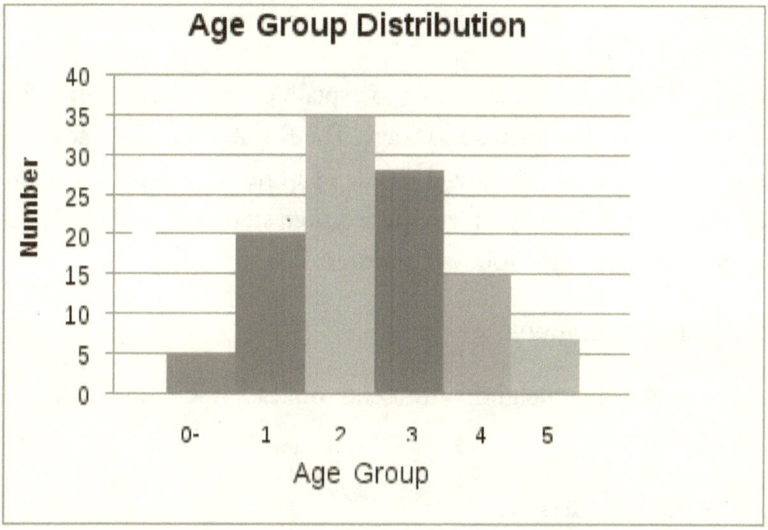
Age Group Distribution
Number
40
35
30
25
20
15
10
5
0
0-
1
2
3
4
5
Age Group

Chart 1.1 – Histogram – Patient Age group Distribution

Throughout this book we are going to explain each statistical tool with examples through R statistical package. The R statistical package is the open source statistical software [1] made up of number of packages and functions which helps to perform the required statistical analysis. It can be downloaded from the following address

https://cran.r-project.org/

Readers can also download the front end tool called R-Studio from the following link

https://www.rstudio.com/

R Studio the front end of the r package is used throughout this book so the readers are requested to download the same before proceeding to the next level.

To fully use the r statistical software the built in packages needs to be downloaded to your system which can be done using the following code in the R studio

install.packages("name of the package")

We will use sample hypothetical sales data set from an online e-commerce site which includes the following variables and all the variables are having numeric values

Day

Custid

Age

Gender (code : Male=1, Female=2)

Productid

Quanity

Price

The sample dataset can be downloaded from the author's website (www.ijsmi.com/dataset.php[1])

installation of packages

install.packages("name of the package")

To import the above mentioned salesdata.xlsx file to r studio environment the following package called "xlsx" to be installed using the below code

```
install.packages("xlsx")
##Next step is to call the same into the r studio environment using
the following code
library("xlsx")
##Setting the working directory wd
wd = "C:\\Users\\xyz\\Documents\\"
setwd(wd)
##Import dataset into r
z <- read.xlsx("salesdata.xlsx",1,check.names=FALSE)
names(z) <- gsub(" ","", names(z))
##Here "check.names=False" helps us to use the names in the xlsx file.
```

Gsub function

We can use $ to refer particular variable or column of a data set

```
age <- z$age
hist(age)
```

1. http://www.ijsmi.com/dataset.php

Frequency table output

	age	Freq
1	15	7
2	16	12
3	17	4
4	18	3
5	19	4
6	20	4
7	21	8
8	22	4
9	23	7
10	24	9
11	25	3
12	26	10
13	27	6
14	28	9
15	29	7
16	30	9
17	31	6
18	32	5
19	33	6
20	34	5
21	35	3
22	36	11
23	37	9
24	38	3
25	39	7
26	40	6
27	41	4
28	42	11
29	43	6
30	44	5

31 45 2
32 46 8
33 47 7
34 48 8
35 49 4
36 50 8
37 51 7
38 52 6
39 53 8
40 54 5
41 55 7
42 56 8
43 57 3
44 58 9
45 59 6
46 60 6
47 61 7
48 62 5
49 63 8
50 64 6
51 65 8
52 66 5
53 67 7
54 68 4
55 69 8
56 70 3

Histogram output

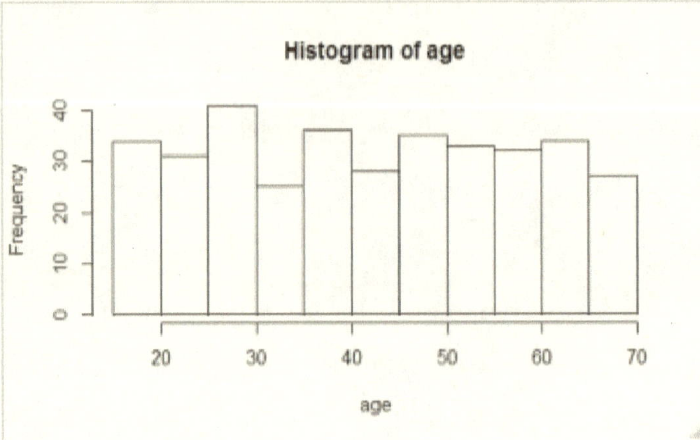

REFERENCES

1. Chalmer, B. J. (2020). Understanding statistics. CRC Press.
2. Gravetter, F. J., Wallnau, L. B., Forzano, L. A. B., & Witnauer, J. E. (2020). Essentials of statistics for the behavioral sciences. Cengage Learning.
3. Rees, D. G. (2018). Essential Statistics for Medical Practice Chapman and Hall/CRC.
4. Momeni, A., Pincus, M., & Libien, J. (2018). Introduction to Statistical Methods in Pathology. Springer International Publishing.
5. Cox, D. R. (2018). Applied statistics-principles and examples. Routledge.
6. Mead, R. (2017). Statistical methods in agriculture and experimental biology. Chapman and Hall/CRC.
7. Pagano, M., & Gauvreau, K. (2018). Principles of biostatistics. Chapman and Hall/CRC.
8. Dixon, W. J., & Massey Frank, J. (1950). Introduction To Statistical Analysis. McGraw-Hill Book Company, Inc; New York.

9. Cook, T. D., & DeMets, D. L. (2007). Introduction to statistical methods for clinical trials. Chapman and Hall/ CRC.
10. Hahs-Vaughn, D. L., & Lomax, R. G. (2013). An introduction to statistical concepts. Routledge.
11. Fisher, R. A. (2006). Statistical methods for research workers. Genesis Publishing Pvt Ltd.
12. Armitage, P., Berry, G., & Matthews, J. N. S. (2008). Statistical methods in medical research. John Wiley & Sons.
13. Stevens, S. S. (1946). On the theory of scales of measurement.
14. Velleman, P. F., & Wilkinson, L. (1993). Nominal, ordinal, interval, and ratio typologies are misleading. The American Statistician, 47(1), 65-72.
15. Stockburger, D. (2001). Introductory statistics: Concepts, models, and applications. Atomic Dog Pub.
16. Hanna, A. R., Rao, C., & Athanasiou, T. (2010). Graphs in statistical analysis. In Key Topics in Surgical Research and Methodology (pp. 441-475). Springer, Berlin, Heidelberg.

Chapter 2 - Descriptive Statistical Tools

Descriptive statistical tools help us to describe the data under consideration.

Measures of Central Tendency

The basic tool but the most widely used tool in statistics is the measures of central tendency. It includes mean, median and mode. The measures of central tendency provide us the vital information about the data set which we are dealing with and the how the dataset can be represented by a single value and around which all other values of the data set are present.

Mean

The most widely used measure of central tendency is mean which is also called as average. It is calculated from all the values of the data.

Advantage of Mean

- It includes all the values of the dataset
- Represents the dataset by a large extent
- Flexible for further mathematical treatment

Disadvantages of mean

- Affected by extreme values
- It is better suited for data measured on ratio scales

Median

Median is calculated after ordering the dataset and it is the middle value of the ordered dataset. If the order set is even number then the average of two middle values is taken as median.

Advantage of Median

- It is not affected by the outliers
- It is suited for the scales of measurements

Disadvantages of median

- Not Flexible for further mathematical treatment

Mode

Mode is the values in the dataset which has highest frequency or highest number of times it appears than the other values in the dataset.

Chapter2 - Sample R code – Descriptive Statistics

install.packages("name of the package")

To import the above mentioned clinical1.xlsx file to r studio environment the following package called "xlsx" to be installed using the below code

```
install.packages("xlsx"
##Next step is to call the same into the r studio environment using the following code
library("xlsx")
##Setting the working directory wd
wd = "C:\\Users\\xyz\\Documents\\"
setwd(wd)
##Import dataset into r
library("xlsx")
z <- read.xlsx("salesdata.xlsx",1,check.names=FALSE)
names(z) <- gsub(" ","", names(z))
##Here "check.names=False" helps us to use the names in the xlsx file. Gsub function
## Using summary function we can get Summary of the variable age in the dataset. We can use $ to refer particular variable or column of a data set
```

Mean function provides the average or mean value of the variable under consideration

```
summary(z$age)
Averagedailysales <-mean(z$quantity)
#To calculate mode table and max function can be used
maxfreq<-max(table(z$age))
mode<-names(sort(-table(z$age)))[1]
```

Measures of Central Tendency output

Min. 1st Qu. Median Mean 3rd Qu. Max.
15.00 28.00 42.00 42.29 56.00 70.00
mode
[1] "16"
Averagedailysales
[1] 2.842697
In the above the measures such as 1st Quarter and 3 Quarter will be discussed in the Measures of Dispersion section

Measures of Dispersion

Measures of dispersion helps us to find out how the values in the data set are spread across the dataset. It helps us to measure how the individual values of the dataset differ from its mean value. Range, Mean Absolute Deviation, Variance and Standard Deviation are the examples of measures of dispersion.

Range

Range is calculated by subtracting minimum value from the maximum value of particular variable.

Mean Absolute Deviation

Mean Absolute Deviation helps us to calculate the average deviation of each value from its mean after ignoring the signs of deviation.

Variance

Variance provides us the valuable information about the spread of data or how much is the deviation for each value in the distribution from the mean of the distribution. In most of the cases the mean and variance together identify the distribution. The unit of measurement of variance is the square of the unit of measurement of the variable under consideration.

Standard Deviation

Standard deviation is the square root of variance which is having the same property as variance and the only difference is the unit of measurement of standard deviation is same as the unit of measurement of the particular variable under consideration

Coefficient of variation

Coefficient of variation is the ratio of standards deviation to the mean of the variable expressed in terms of percentage.

Standard Error

Standard error is calculated by dividing the standard deviation by square root of the dataset size (n).

Measures of Dispersion

```
##Import dataset into r

library("xlsx")

z <- read.xlsx("salesdata.xlsx",1,check.names=FALSE)

names(z) <- gsub(" ","", names(z))
```

##Here “check.names=False” helps us to use the names in the xlsx file. Gsub function

Using summary function we can get Summary of the variable age in the dataset. We can use $ to refer particular variable or column of a data set

```
summary(z$age)
```

we can use psych to get the measures of dispersion apart from summary function

```
library(psych)

describe(z$age)

#We can use boxplot function to plot the Boxplot in r

boxplot(z$age)
```

Measures of Dispersion output

mean	median	trimmed
42.29	42	42.34

Sd	mad	min	max
16	20.76	15	70

skew	kurtosis
0	-1.2

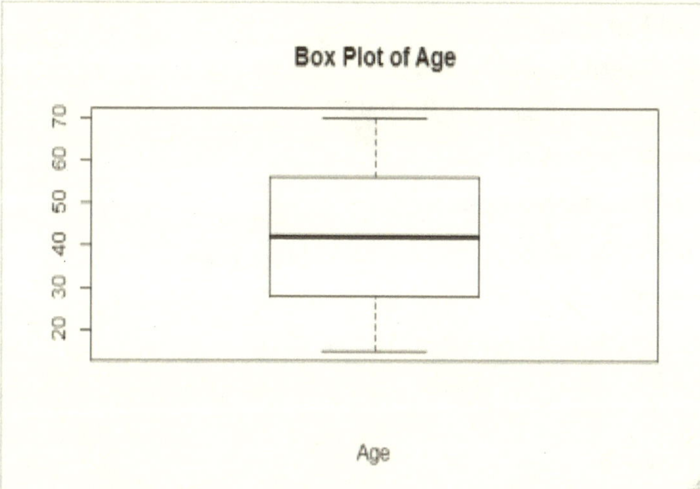

In the above output, the measures of such as Skewness and kurtosis will be discussed in the Measures of Shape section

Measures of Location (Quartiles and Percentiles)

Measures of location include Quartiles and Percentiles. Percentiles divide the dataset into hundreds and the quartiles divide the given dataset into quarters (Q1, Q2, Q3, Q4).

Percentile

Percentile calculated as location measure and 95th percentile means the 95% of the dataset values are less than that particular value. It is mostly used as a relative measure such as student scores in a test. If one student got 98th percentile means his score is more than or equal to 98% percent of the students who scored mark in the particular test.

Interquartile Range

Interquartile range is calculated by subtracting the value of first Quartile (Q1) from third Quartile (Q3) i.e. Q3-Q1

Measures of Location

Quartiles<-quantile(age)
Percentiles<-quantile(age,c(0.1,0.2,0.95))
Interquartilerange <-IQR(age)

Measures of Location output

0% 25% 50% 75% 100%
15 28 42 56 70
Percentiles
10% 20% 95%
21 26 67
Interquartilerange
[1] 28

Measures of Shape (Skewness and Kurtosis)

Measures of Skewness and Kurtosis help to find out shape of the distribution of the dataset

Skewness

Skewness indicates the direction of the distribution curve. It helps us to find whether the distribution has a peak in left side of the distribution curve (positively skewed) or middle (no Skewness) or right side of the curve (negatively skewed).

Kurtosis

Kurtosis indicates the height of the distribution curve. If it has a high peak (leptokurtic or high positive kurtosis value) or normal peak (mesokurtic or kurtosis is 0) or no peak / flat (platykurtic or high negative kurtosis value)

Measures of Shape

##Import dataset into r

library("xlsx")

z <- read.xlsx("salesdata.xlsx",1,check.names=FALSE)

names(z) <- gsub(" ","", names(z))

##Here "check.names=False" helps us to use the names in the xlsx file. Gsub function

Using summary function we can get Summary of the variable age in the dataset. We can use $ to refer particular variable or column of a data set

summary(z$age)

we can use psych to get the measures of dispersion apart from summary function

library(psych)

describe(z$age)

Measures of Shape output

mean median trimmed

42.29 42 42.34

Sd mad min max range

16 20.76 15 70 55

se

0.85

skew kurtosis

0 -1.2

References

1. Briggs Baffoe-Djan, J., & Smith, S. A. (2020). Descriptive statistics in data analysis. The Routledge handbook of research methods in applied linguistics.
2. Rohana, R., & Ningsih, Y. L. (2020). Students'statistical reasoning in statistics method course. Jurnal Pendidikan Matematika, 14(1), 81-90.
3. Maity, R. (2018). Basic Statistical Properties of Data. In Statistical Methods in Hydrology and Hydroclimatology (pp. 53-92). Springer, Singapore.
4. Kaur, P., Stoltzfus, J., & Yellapu, V. (2018). Descriptive statistics. International Journal of Academic Medicine, 4(1),

60.

5. Rodrigues, C. F. D. S., Lima, F. J. C. D., & Barbosa, F. T. (2017). Importance of using basic statistics adequately in clinical research. Revista brasileira de anestesiologia, 67(6), 619-625.
6. Garrocho-Rangel, J. A., Ruiz-Rodríguez, M. S., & Pozos-Guillén, A. J. (2017). Fundamentals in Biostatistics for Research in Pediatric Dentistry: Part I–Basic Concepts. Journal of Clinical Pediatric Dentistry, 41(2), 87-94.
7. R Core Team (2018). R: A language and environment for statistical computing. R Foundation for Statistical Computing, Vienna, Austria. URL https://www.R-project.org/.
8. https://www.rstudio.com/
9. https://cran.r-project.org/web/packages/rio/rio.pdf
10. Komsta, L., & Novomestky, F. (2015). Moments, cumulants, skewness, kurtosis and related tests. R package version 0.14.
11. Wilcox, R. R., & Keselman, H. J. (2003). Modern robust data analysis methods: measures of central tendency. Psychological methods, 8(3), 254.
12. Jaggi, S. (2003). Descriptive statistics and exploratory data analysis. Indian Agricultural Statistics Research Institute, 1, 1-18.
13. Kvanli, A. H., Pavur, R. J., & Guynes, C. S. (2000). Introduction to business statistics. Cincinnati, OH: South-Western.
14. Fisher, M. J., & Marshall, A. P. (2009). Understanding descriptive statistics. Australian Critical Care, 22(2), 93-97.
15. Weisberg, H., & Weisberg, H. F. (1992). Central tendency and variability (No. 83). Sage.
16. Zar, J. H. (1999). Biostatistical analysis. Pearson Education India.

17. Larson, M. G. (2006). Descriptive statistics and graphical displays. Circulation, 114(1), 76-81.
18. Fisher, M. J., & Marshall, A. P. (2009). Understanding descriptive statistics. Australian Critical Care, 22(2), 93-97.
19. McHugh, M. L., & Hudson-Barr, D. (2003). Descriptive statistics, part II: Most commonly used descriptive statistics. Journal for Specialists in Pediatric Nursing, 8(3), 111-116.
20. Munro, B. H. (2005). Statistical methods for health care research (Vol. 1). lippincott williams & wilkins.
21. Altman, D. G. (1990). Practical statistics for medical research. CRC press.
22. Marshall, G., & Jonker, L. (2010). An introduction to descriptive statistics: A review and practical guide. Radiography, 16(4), e1-e7.
23. Weiss, N. A., & Weiss, C. A. (2012). Introductory statistics. London: Pearson Education.
24. Sonnad, S. S. (2002). Describing data: statistical and graphical methods. Radiology, 225(3), 622-628.
25. Woolson, R. F., & Clarke, W. R. (2011). Statistical methods for the analysis of biomedical data (Vol. 371). John Wiley & Sons.
26. Doane, D. P., & Seward, L. E. (2011). Measuring skewness: a forgotten statistic?. Journal of Statistics Education, 19(2).

Chapter 3 - Probability and Probability Distributions

Introduction

Probability is defined as the chance or likelihood of happening of event or outcome (result from an experiment or activity).

Sample space and events

Sample space in probability is defined as the set of all possible outcomes. Event is the subset of sample space. In tossing an unbiased coin, (Head and Tail) is the sample space. In the event of patient survival status, the sample space is (alive or dead).

Classical Probability

The classical probability is useful when all the outcome of event is equally likely. In tossing coin, head and tail outcomes are equally likely & mutually exclusive and is calculated using relative frequency of occurrence of an outcome i.e. by dividing the number of times the outcome under consideration can occur(say getting a head) vs total number of times the event is repeated(number of times the coin tossed). In classical probability, the possible number of times of event can be calculated even before event is conducted. Classical probability can be calculated only when sample space is finite.

EMPIRICAL PROBABILITY

Empirical probability is calculated using relative frequency of an outcome of event when similar event is conducted in the past and it is calculated as number of times the desired outcome occurred vs total event conducted in the past.

Properties of Probability

- Probability of an outcome can take a value between 0 and 1.
- Probability of all possible outcome of an event sums to 1

Independent event

If two events A and B are said to be independent if the occurrence of an event (A) does not dependent on the occurrence of another event (B).

Conditional Probability

Conditional probability indicates the probability of happening of an event (B) depends on the probability of happening of another event (A). If probability of a Customer who is having a credit card is (A) then the probability of customer buying a smart phone that he is having credit is given by

P (B/A) = P (A and B)/ P (A)

Where P (A and B) is probability of patient having credit card buy smart phone

---

BAYES THEOREM

Let us take the reverse of the above example i.e. probability of patient having credit card (A) given that he bought smart phone (B)

P (A/B) = P (A and B)/ P (B)

= [P (B/A). P (A)] / [(P (A) . P (B/A)) . (P(A) P(B/A)]

Probability Distribution

Random variable

Random variable is a variable whose values are determined by a chance or random factors. Discrete random variables take integer values or whole numbers whereas the continuous random variable can take any values between two integers or interval or real numbers.

Probability distributions

Discrete Probability distribution of a discrete random variable defines all possible values the random variable can take along with the respective probability for each value. Binomial and Poisson are the common discrete distributions used frequently.

Continuous Probability distribution of continuous random variables defines probability as the area under the curve between given interval (say a, b). Normal distribution is the most frequently used continuous distribution.

BINOMIAL PROBABILITY distribution

A random variable is said to follow binomial distribution when it takes only two outcomes say "Success" or "Failure". The outcomes are independent of each trail or experiment and have the same probability of success throughout the experiments or trials. If we want to find out the probability of 3 customers out of 20 persons buys smart phone when they clicked the add to cart link in an online site when the probability of a customer buys smart phone is 0.3 when clicked the add to cart link, then Binomial distribution can be used to find out the probability as the customers are independent and having a same probability of success.

Poison Probability distribution

A random variable said to follow the poison distribution if it takes positives values from 0 to infinity within the given interval of time or space. The interval of time might be one minute or one hour and interval of space may be a human cell or piece of cloth. Poison distribution is useful for rare events. For example, if x is the number of customers arriving at bank within one hour then it can take values such as 0, 1, 2....., but the value cannot be negative (-1). In this case, if 5 is average number of customers arriving at the bank then we can use poison distribution to calculate the probability of 10 arriving at the bank within one hour.

Normal Probability distribution

A random variable said to follow the normal distribution if its density curve is bell shaped curve and the curve is symmetric in nature. The random variable which follows normal distribution has 67% of its values lie between mean ± Standard deviation, 95% of its values lies between its mean ± 2.standard deviation and 99% of its values lies between its mean ± 3.standard deviations. Most important property of normal distribution is almost all the distribution follows normal distribution when the sample size is large.

Binomial distribution example

Let us consider a problem of finding the probability of 3 customers Out of 20 buying the smart phone after clicking the add to cart link in an online site successfully when the overall success rate is given as 30% . In this case we can use binomial function in R to calculate the probability of success (Buying) for 3 customers

Probability distributions

```
x<-3
binomialprobbability<- dbinom(x, size=20, prob=0.3)
binomialprobbability
success <- 0:20
Plot(success, dbinom(success, size=20, prob=.3),type='h')
```

Probability distributions output

```
binomialprobbability
[1] 0.07160367
```

THE PLOT FUNCTION HELPS to draw Binomial Probability distribution of number of success (number of customers buying out of 20) ranging from 0:20 when the probability of success or success rate is 30%

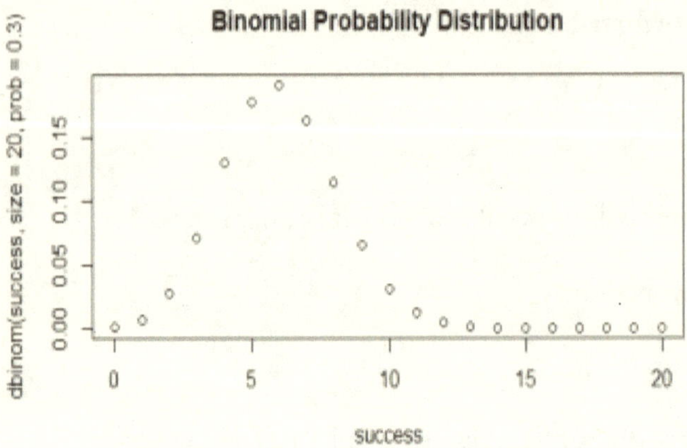

Poison distribution example

Let us consider a problem of finding the probability of 10 customers arriving at a bank within one hour when the average number of customers arriving at a bank is 5 per hour. The probability can be found using poison function in R

Probability distributions

```
x=10

Poisonprobability<-dpois(x,5)

Poisonprobability

customers <- 0:10

plot(customers, dpois(customers,5),

main = "Poison Probability Distribution")
```

Probability distributions output

```
Poisonprobability

0.01813279
```

The plot function helps to draw Poison Probability distribution of number of customers arriving at a bank ranges from 0:20 when the average number of customers is 5 per hour

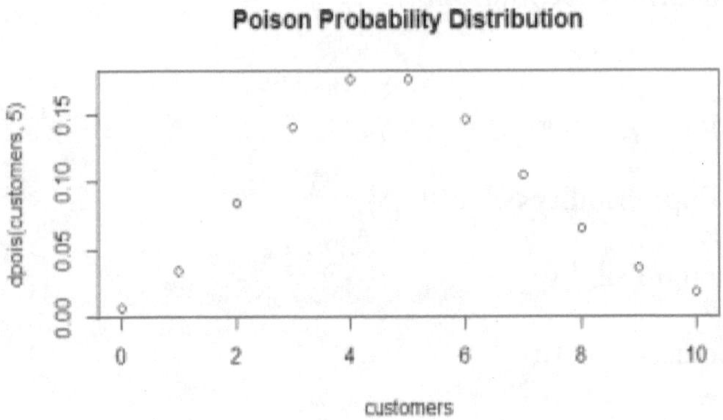

Normal distribution example

Let us consider a problem of finding the probability of 10 products sold per day in e commerce website when generally on an average 7 products are sold with standard deviation of 2. To calculate the probability we use the normal function in R

Probability distributions

```
x1<-10

normalproability<-dnorm(x1, mean=7, sd=2)

prd<-0:25

plot(prd, dnorm(prd, mean=7, sd=2),

main = "Normal Probability Distribution")
```

Probability distributions output

normalprobability

[1] 0.0647588

The plot function helps to draw Normal Probability distribution of number of products sold per day ranges from 0:25 when the average product sold per day is 7 with standard deviation 2

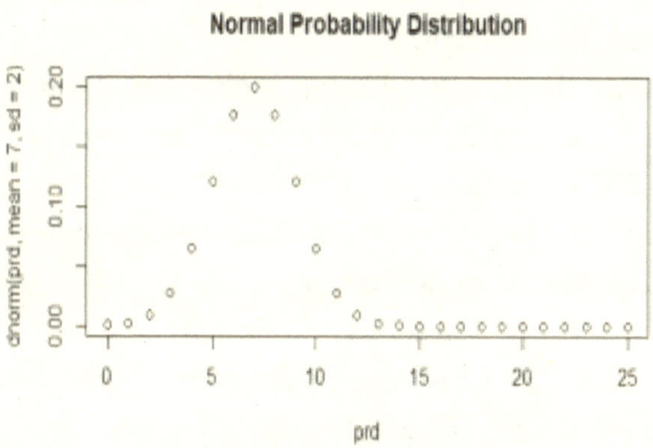

REFERENCES

1. Blitzstein, J. K., & Hwang, J. (2019). Introduction to probability. Crc Press.
2. Singh, A. K., Gewali, L. P., & Khatiwada, J. (2019). New Measures of Skewness of a Probability Distribution. Open

Journal of Statistics, 9(5), 601-621.

3. R Core Team (2018). R: A language and environment for statistical computing. R Foundation for Statistical Computing, Vienna, Austria. URL https://www.R-project.org/.
4. https://www.rstudio.com/
5. Tuckwell, H. C. (2018). Elementary applications of probability theory. Routledge.
6. Chatfield, C. (2018). Statistics for technology: a course in applied statistics. Routledge.
7. Di Paola, G., Bertani, A., De Monte, L., & Tuzzolino, F. (2018). A brief introduction to probability. Journal of thoracic disease, 10(2), 1129.
8. Grech, V., & Calleja, N. (2018). WASP (Write a Scientific Paper): Probability–Poisson and binomial distributions. Early human development.
9. Forsyth, D. (2018). Useful Probability Distributions. In Probability and Statistics for Computer Science (pp. 115-137). Springer, Cham.
10. Tuckwell, H. C. (2018). Elementary applications of probability theory. Routledge.
11. Leemis, L. M., Luckett, D. J., Powell, A. G., & Vermeer, P. E. (2017). Univariate Probability Distributions. In Computational Probability Applications (pp. 133-147). Springer, Cham.
12. Sokal, R. R., & Rohlf, F. J. (1969). The principles and practice of statistics in biological research (pp. 399-400). San Francisco:: WH Freeman and company.
13. Rohatgi, V. K., & Saleh, A. M. E. (2015). An introduction to probability and statistics. John Wiley & Sons.
14. Ross, S. M. (2014). Introduction to probability models. Academic press.

15. Schervish, M. J., & DeGroot, M. H. (2014). Probability and statistics. Pearson Education.

Chapter 4 - Sampling

Sampling concepts are very essentials in any type of quantitative research involving data collection. Sample is a subset of the population under study. The researchers normally generalize the findings from the sample to the whole population. The quantities estimated from sampling process are called sample statistics and the related population quantities are called population parameters.

Advantages of sampling

- Sampling helps us to save time and cost
- Sampling will be the only option when we conduct destructive kind of study (To estimate the life time of bulbs etc.)
- Sampling reduces the non-sampling errors

Disadvantages of sampling

- Results might not be generalizable to the fuller extent
- Errors might creep in due to sampling methodology and sample size results in sampling error

Types of sampling

Probability sampling

Probability sampling method includes selecting the sample randomly from the population so as to ensure that it is a representative of the population from which it is selected. It also ensures each population unit has the equal probability of being included in the sample. There are different types of probability sampling methods available which are given below

Simple Random Sampling

If want to study the customer buying behavior of a particular product from a hypermarket we should have the list of buyers who have bought the product. We can randomly select the customer IDs from the list using any random generators. It ensures each customer has equal chance of being included in the sample and the results can be generalized to the entire population.

Stratified Random Sampling

Stratified random sampling method enables us to give weight to the different categories of the population according to their size in the population. Stratified Random sampling is useful when the population is heterogeneous. Heterogeneous population is divided into homogeneous stratum or categories. For example, If want to include the buyers in the sample according to the age group customers then each of the categories such as young, middle and old age group will have representation in the sample according to their size in the population. The population is first divided into stratum or category and sample unit will be selected from each stratum and number of units selected from each stratum depends on the size of the stratum.

Systematic Random Sampling

Systematic random sampling is also similar to simple random sampling except the fact that every kth unit of the sampling list will be included into the sample. The number k is determined by dividing the population size (N) divided by the sample size (n). The sampling process starts with selecting a sampling unit within the first k units of the sampling list and after that every kth is selected till the sample size is reached.

Cluster Sampling

In Cluster sampling, the population units are grouped to form natural clusters and sample clusters are selected from the list of all clusters and sampling units in each cluster is included in the sample. For example, cluster of customers are formed based on the preference (price conscious, quality conscious etc.) and then the sample clusters are selected and each patient from the cluster is included in the sample.

Multistage sampling

If the population size is high and if it is difficult to get the list of population then multistage sampling can be used wherein samples will be taken at different stages of the population before reaching the actual population unit. For example if want to take sample of customers of product from Chain of a US Supermarket then at the first stage the sample states will be selected from the states list, then in the second stage sample counties will be selected and in the third stage supermarkets will be selected and finally the customers from that particular supermarket will be selected at random.

Nonprobability sampling

Nonprobability sampling involves selecting the samples from the population based on the researcher's convenience in terms of time and cost. The sample which is taken from the population is not a representative of the population and results are not generalizable. Quota sampling, snow ball sampling and convenient sampling are the examples of nonprobability sampling methods.

Quota Sampling

Quota sampling method involves first dividing the population into categories and selecting the sample units at the convenience. Here the actual proportion of each category is not taken into consideration and only the categories are considered before the sample is taken.

Snowball Sampling

Snowball sampling involves selecting the first sampling unit and then asks the selected sampling unit to recommend another sampling unit till the required sample size is reached. Snowball sampling is useful when the sampling unit under study is rare and not easily approachable.

Convenient Sampling

Convenient sampling involves selecting the sample based on the researcher's convenience in in order to save time and cost.

Errors in Sampling

Sampling Error

Sampling error is the error introduced in the sample estimates due to sampling process. It reflects the deviation of estimates of sample statistics from its corresponding population parameters.

Non sampling error

Non sampling errors are those errors which arise due to non-response, errors in measurement instruments, interviewer bias, respondent errors and errors at data entry stage.

Non response error

Non response error affects the accuracy of the sampling estimates. The reason for non-response to be clarified by the researcher whether the non-response is due to any random cause or any specific reason or pattern. One of the way to manage the non-response rate, it is advisable to increase the size of the sample in proportion to projected non-response rate

Errors in measurement instruments

Errors in data collection instruments such as customer data collection forms, survey forms in terms of wrong questions, wrong options to the questions Interviewer, respondent bias and errors at the data entry stage

Errors creep in due to interviewer and respondent bias such as not understanding the question, deliberately providing wrong answers. Errors might also creep in at the time of entering the data.

Sample size determination

Sample size determination is one of the most important phases of sampling process. It helps us to achieve the desired accuracy of the parameters being estimated and the size of the sample impacts accuracy of the sample estimates. It is directly proportional to sampling error. When the sample size increases the sampling error decreases. The sample size determination depends on the following which will be covered in the following chapters

Main Outcome variable

Level of significance

Level of significance is the probability of observing a given value by chance when null hypothesis is true. It is called Type I error (probability of rejecting null hypothesis when it is actually true) in the hypothesis testing which will be discussed in detailed later the hypothesis testing section.

Power of test

The power test is probability of accepting the null hypothesis when it is actually wrong and will be discussed in detailed later in the hypothesis testing section.

Central Limit Theorem

Central limit theorem is an important theorem which states that when we draw number of samples with sufficiently large size (n>30) from a population with mean μ and standard deviation σ, then the sample means calculated from the samples follow normal distribution.

References

1. Shackleton, N., Von Randow, M. A. R. T. I. N., & Greaves, L. (2019). Methods of Sampling in Quantitative Health Research. Researching Health: Qualitative, Quantitative and Mixed Methods, 227.
2. Williams, B. K., & Brown, E. D. (2019). Sampling and analysis frameworks for inference in ecology. Methods in Ecology and Evolution, 10(11), 1832-1842.
3. Lynn, P. (2019). The advantage and disadvantage of implicitly stratified sampling. methods, data, analyses, 13(2), 14.
4. Cevallos-Torres, Lorenzo, and Miguel Botto-Tobar. "Process sampling." Problem-Based Learning: A Didactic Strategy in the Teaching of System Simulation. Springer, Cham, 2019. 13-31.
5. Cochran, W. G. (2007). Sampling techniques. John Wiley & Sons.
6. Acharya, A. S., Prakash, A., Saxena, P., & Nigam, A. (2013). Sampling: Why and how of it. Indian Journal of Medical Specialities, 4(2), 330-333.
7. Henry, G. T. (1990). Practical sampling (Vol. 21). Sage.
8. Kothari, C. R. (2004). Research methodology: Methods and techniques. New Age International.
9. Levy, P. S., & Lemeshow, S. (2013). Sampling of populations: methods and applications. John Wiley & Sons.
10. Madow, W. G. (1949). On the theory of systematic sampling, II. The Annals of Mathematical Statistics, 20(3), 333-354.
11. Bellhouse, D. R. (1988). 6 Systematic sampling. Handbook of

statistics, 6, 125-145.

12. Kandola, D., Banner, D., O'Keefe-McCarthy, S., & Jassal, D. (2014). Sampling Methods in Cardiovascular Nursing Research: An Overview. Canadian Journal of Cardiovascular Nursing, 24(3).
13. Sukhatme, P. V. (1957). Sampling theory of surveys with applications. The Indian Society Of Agricultural Statistics; New Delhi.
14. Hansen, M. H., & Hurwitz, W. N. (1953). Sample survey methods and theory. Vol. I. John Wiley And Sons, Inc.; New York.
15. Sudman, S. (1976). Applied sampling (No. 04; HN29, S8.). New York: Academic Press.
16. Som, R. K. (1973). A manual of sampling techniques. A manual of sampling techniques.
17. Gravetter, F. J., & Forzano, L. A. B. (2018). *Research methods for the behavioral sciences*. Cengage Learning.
18. Kahn, H. A., & Sempos, C. T. (1989). Statistical methods in epidemiology (Vol. 12). Monographs in Epidemiology & B.
19. Chow, S. C., Shao, J., Wang, H., & Lokhnygina, Y. (2017). Sample size calculations in clinical research. Chapman and Hall/CRC.
20. Zhang, X., & Zhao, N. (2017). Sample size estimation for survival analysis in single-arm trials. Fudan University Journal of Medical Sciences, 44(4), 517-520.
21. Eng, J. (2003). Sample size estimation: how many individuals should be studied?. Radiology, 227(2), 309-313.
22. Suresh, K. P., & Chandrashekara, S. (2012). Sample size estimation and power analysis for clinical research studies. Journal of human reproductive sciences, 5(1), 7.
23. Jones, S. R., Carley, S., & Harrison, M. (2003). An introduction to power and sample size estimation. Emergency

Medicine Journal, 20(5), 453-458.

Chapter 5 - Estimation and Statistical Inference

Estimation and statistical inference provides us valuable information about the population parameters through sample statistics. Estimation helps to find out the value of the unknown parameter through the sample statistics whereas the statistical inference helps to us test whether a particular value for the population parameters is true or not. The estimation process can be divided into point and interval estimation which are discussed below.

Point Estimation

Point estimation provides us a single value for the unknown population parameters. Maximum likelihood estimation and method of moments and method of least squares are the methods available for estimating the population parameters from the sample.

Maximum Likelihood estimation method

If we want to estimate the proportion of persons who have LDL cholesterol more than 160mg/dl from New York City, it is not possible to test all the persons and get the proportion. Hence we need to take sample and estimate the unknown population proportion in the New York City. Maximum likelihood estimation will give you an estimator of the population which will maximize the likelihood of observing particular cholesterol level of patients in the sample.

For example, in this case if x is a random variable which takes the value 1 if the cholesterol level is higher than 160mg/dl and 0 if it is less. If take a sample of size 100 and then sample data will be (1, 0, 1 , 0 ,1 ,0.....1). Here let us say sample proportion of persons (sample mean proportion $\bar{p}$) who are having cholesterol level more than 160mg/dl is 0.6 or 60% then $\bar{p}$ is an estimator of the unknown population proportion $\bar{P}$ - Population mean proportion) as it maximizes the likelihood or probability of observing the sample data ((1, 0, 1 , 0 ,1 ,0.....1) .

Here $\bar{p}$ is the maximum likelihood estimator and 0.6 is the maximum likelihood estimate (the value the estimator takes)

Method of Moments

In the method of moments we need to equate the population moments with sample moments through the sample data.

Method of least squares

Method of least squares estimates the population parameter from sample by minimizing the sum of square of deviation of the sample values from the sample statistics (mean).

Confidence interval

Interval estimation provides an interval where the population parameter lies with a certain amount of confidence. Care to be taken when we interpret the Confidence intervals. When we construct a 95% confidence interval for a population parameter from the sample, then if repeat the process with different samples from the same population several times, then 95% of total intervals will contain the unknown population parameter.

Confidence interval for population mean

To calculate the confidence interval for population mean, we need calculate the following

- Sample mean

- Standard error of mean

The theoretical distribution values of the variable at (1- (level of significance/2)) *100.

For example, we want to calculate the mean age of the customer who bought a particular product from an organization in New York City and we have the data from a sample of 100 customers interviewed. If the age variable follows normal distribution then the theoretical distribution value standard normal variable (z) to be find out from the normal table at (1- (level of significance 5%) /2) i.e. (1-0.025)=0.975. The theoretical value of z from the normal distribution can found as 1.96.

Confidence interval for mean age of customer of the organization = sample mean age ± 1.96* (sample standard deviation/square root of n)

Let us consider the salesdata.xlsx data set for carrying out estimation and hypothesis testing. It can be downloaded from author's website www.ijsmi.com/dataset.php[1]

The following is the structure is data set

- custId
- day
- age
- gender (coded as Male=1, Female=2)
- productid (coded as 1 to 15)
- incomerange(coded as 1 to 3)
- quantity
- price
- rating (coded as 1 to 5)
- Recommend(Codes as 0 or 1)

1. http://www.ijsmi.com/dataset.php

Confidence interval for population mean

```
library(“xlsx”)
##Import dataset into r
z <- read.xlsx("salesdata.xlsx",1,check.names=FALSE)
names(z) <- gsub(" ","", names(z))
#Define a function to calculate the confidence interval
confint = function(x,x2,x3)list(
#mean
Mean = mean(x),
#Standard error
SE = sd(x)/(sqrt(x2)),
#Theoretical distribution value of the variable
theoretical = qnorm(x3),
#upper and lower limit
cupper = (mean(x) + (qnorm(x3) * (sd(x)/(sqrt(x2))))),
clower = (mean(x) - (qnorm(x3) * (sd(x)/(sqrt(x2)))))
)
#Sample size or count of variable age number of customers
s <- length(age)
#1-level of significance
s2 <- (1-(0.05/2))
#Call the function using the variable, sample size and 1-level of
significance
confint(age,s,s2)
```

Confidence interval for population mean output

$Mean
[1] 42.29494
$SE
[1] 0.8482503
$theoretical
[1] 1.959964
$cupper
[1] 43.95748
$clower
[1] 40.6324

Confidence interval for difference in two population means

To calculate the confidence interval for difference in two population means, we need to calculate the following

Difference in Sample means

Standard error of means

The theoretical distribution values of the variable at (1- (level of significance/2)) *100.

Confidence interval for difference between two population means

```
library("xlsx")
##Import dataset into r
z <- read.xlsx("salesdata.xlsx",1,check.names=FALSE)
names(z) <- gsub(" ","", names(z))
#Define a function to calculate the confidence interval
confintdm <- function(x,y,x2,y2,x3)
{
# two sample means
M1<- mean(x)
M2<-mean(y)
#pooled standard deviation
PS1 <- sqrt((((x2-1)*sd(x))+((y2-1)*sd(y)))/(x2+y2-1))
#Standard error
SE <-PS1 * sqrt((1/x2)+(1/y2))
#Theoretical distribution value of the variable
tv <- qnorm(x3)
#upper and lower limit
cupper <- ((M1-M2) + (tv * SE))
clower <- ((M1-M2)+(tv*SE))
print(PS1)
print(cupper)
print(clower)
}
#Sample size or count of variable age number of patients
gen<-z$gender
agef<- age[gen==2]
agem<- age[gen==1]
sf <- length(agef)
sm<-length(agem)
#1-level of significance
s2 <- (1-(0.05/2))
```

#Call the function using the variable, sample size and 1-level of significance
Confintdm(agef,agem,sf,sm,s2)

Confidence interval for two population means output

```
confintdm(agef,agem,sf,sm,s2)
[1] 3.982612
[1] -3.014709
[1] -3.014709
```

Confidence interval for population proportion

To calculate the confidence interval for population proportion, we need to calculate the following

- Sample proportion
- Standard error of proportion

The theoretical distribution values of the variable at (1- (level of significance/2)) *100.

Confidence interval for population proportion

```
library("xlsx")
##Import dataset into r
z <- read.xlsx("salesdata.xlsx",1,check.names=FALSE)
names(z) <- gsub(" ","", names(z))
#Define a function to calculate the confidence interval
confintp <- function(x,x2,x3)
{
#proportion
P<- mean(x)
#Standard error
SE <- sqrt((P*(1-P))/x2)
#Theoretical distribution value of the variable
tv <- qnorm(x3)
#upper and lower limit
cupper <- (P + (tv * SE))
clower <- (P+(tv*SE))
print(P)
print(cupper)
print(clower)
}
s <- length(z$recommend)
#1-level of significance
s2 <- (1-(0.05/2))
#Call the function using the variable, sample size and 1-level of significance
confintp(z$recommend,s,s2)
```

Confidence interval for population proportion output

```
confintp(z$Recommend,s,s2)
[1] 0.5337079
[1] 0.5855286
[1] 0.5855286
```

Confidence interval for difference between two population proportions

To calculate the confidence interval for difference between two population proportions, we need to calculate the following

- Sample proportion
- Standard error of proportion

The theoretical distribution values of the variable at (1- (level of significance/2)) *100.

Confidence interval for difference between two population proportion

```
#Define a function to calculate the confidence interval
confintdp <- function(x,y,x2,y2,x3)
{
#proportion
P1<- mean(x)
P2<- mean(y)
#Standard error
SE <- sqrt(((P1*(1-P1))/x2)+ (P2*(1-P2))/y2)
#Theoretical distribution value of the variable
tv <- qnorm(x3)
#upper and lower limit
cupper <- ((P1-P2) + (tv * SE))
clower <- ((P1-P2) - (tv*SE))
print(P1)
print(P2)
print(cupper)
print(clower)
}
#Sample size or count of variable age number of patients
rf <- z$Recommend[gen == 2]
rm <- z$Recommend [gen == 1]
sf <- length(survivalf)
sm <- length(survivalm)
#1-level of significance
s2 <- (1-(0.05/2))
#Call the function using the variable, sample size and 1-level of significance
confintdp(survivalf,survivalm,sf,sm,s2)
```

Confidence interval for difference between two population proportion output

```
confintdp(rf,rm,sf,sm,s2)
[1] 0.5511364
[1] 0.5166667
[1] 0.1380505
[1] -0.06911107
```

Hypothesis testing

Hypothesis testing involves testing value for the unknown population parameter from the given sample data. Generally Hypothesis testing involves the following steps

Steps in Hypothesis testing

Defining of Null Hypothesis and Alternative hypothesis

The null hypothesis as the name suggest is designed to take a neutral view on the parameter being tested against the alternative hypothesis. Alternative hypothesis is the one the researcher wants to test on the parameter most of the times.

If company wants to test whether the marketing campaign offering discount during the festive season has increased the average number of visitors to their website or not

Then in this case the **null hypothesis** (denoted by Ho) will be defined as

Ho: There is no difference between pre and post campaign average visitors to website

The **alternative hypothesis** (denoted Ha) will be defined as

Ha: Average number of visitors increased after the marketing campaign

Define level of significance and power of test

The next step in the hypothesis testing is to decide how much error can be permissible while taking a decision as the decision is made through sample statistics. The errors in the hypothesis testing is divided into Type I and Type II as in the below table

	Actual Status of Null Hypothesis – True	**Actual Status of Null Hypothesis - False**
Status of Null Hypothesis from Hypothesis testing - True	Right Decision	Type II Error (β)
Status of Null Hypothesis from Hypothesis testing - False	Type I error (α)	Right Decision

Here the Type I error is serious than the Type II error and it will maintained at lower level than the Type II error. The reason for the same is if wrongly reject the null hypothesis means we are accepting a situation which worse than the current one and the other hand if wrongly accept the hypothesis then we are losing one opportunity.

The type I error (α) is also known as Level of Significance and 1- (β) is known as the power of the test. Power of the test helps us to detect the true effect of the test in rejecting the null hypothesis when it is actually wrong under the given level of significance.

Compute test statistics

To compute the test statistics the following is required generally

- The Sample Statistics at the level of Null hypothesis and/or alternative hypothesis
- Standard error

Level of significance

Compare p values with level of significance or calculate theoretical statistics and compare it calculated statistics

The theoretical value of the test statistics needs to be computed from theoretical distribution using the degrees of freedom and level of significance. The degrees of freedom for a test statistics is defined as the total sample size minus the number of parameters which will be estimated from the sample itself. If the theoretical value is greater than the computed value of the test statistics then the test statistics observed from the data is due to chance only and we proceed to accept the null hypothesis.

After the use of statistical software in data analysis, the concept of p values are used and computed from the test statistics. P value is defined as the probability of observing the test statistics based on the chance alone. Normally it is compared with the level of significance. If it is greater than the level of significance we can conclude that the test statistics observed is due to chance only and proceed to accept the null hypothesis.

Draw inference and conclusion

Final stage of hypothesis testing is to either accept or reject the null hypothesis at the given level of significance and draw conclusion

Testing sample mean vs. population mean (z – test for large samples)

Hypothesis Testing sample mean vs. population mean– c8

```
library("xlsx")
##Import dataset into r
z <- read.xlsx("salesdata.xlsx",1,check.names=FALSE)
names(z) <- gsub(" ","", names(z))
#Hypothesis testing sample mean vs. population mean
#We would like to test the mean age of the customers in the sample
dataset as 45
#H0 : μ = 45
#H1 : μ <> 45
x1<-mean(age)
x2<-sd(age)
z1<-(x1-45)/(x2/sqrt(length(age)))
p1<-2 * pnorm(- abs(z1))
cat("Mean Age = ",x1,"\n")
cat("Stardard Deviation =",x2,"\n")
cat("p value=",p1,"\n")
```

Hypothesis Testing sample mean vs. population mean output

```
Mean Age = 42.29494
Stardard Deviation = 16.00476
p value= 0.001427741
```

Inference: As the p value is less than 0.05(0.0014) we reject the null hypothesis that the mean is equal to 45 accept the alternative hypothesis H1 that the mean is not equal to 45

Testing difference between two sample means(z – test for large samples)

Hypothesis Testing difference between two population means

```
library("xlsx")
##Import dataset into r
z <- read.xlsx("salesdata.xlsx",1,check.names=FALSE)
names(z) <- gsub(" ","", names(z))
#Hypothesis testing two means
#We would like to test the whether there is any significance
difference between the #mean age of male and female customers in
sample salesdata dataset
#H0 : meanagemale = meanagefemale
#H1 : meanagemale <> meanagefemale
age<-z$age
gen<-z$gender
agef<- age[gen==2]
agem<- age[gen==1]
m1 <- mean(agef)
m2 <- mean(agem)
sdm1 <- sd(agef)
sdm2 <- sd(agem)
nm1 <- length(agef)
nm2 <- length(agem)
zm1 <- (m1-m2)/(sqrt((sdm1*sdm1)/nm1)+((sdm2*sdm2)/nm2))
pm1 <- 2*pnorm(-abs(zm1))
cat("Mean Age of Male = ",m2,"\n")
cat("Mean Age of Female =",m1,"\n")
cat("Standard deviation of Age of Male = ",sdm2,"\n")
cat("Standard deviation of Age of Female =",sdm1,"\n")
cat("p value=",pm1,"\n")
```

Hypothesis Testing difference between two population means output

cat("Mean Age of Male = ",m2,"\n")
Mean Age of Male = 44.19444
Mean Age of Female = 40.35227
Standard deviation of Age of Male = 15.50567
Standard deviation of Age of Female = 16.31549
value= 0.1342329

INFERENCE: AS THE P value is greater than 0.05(0.134) we accept the null hypothesis that the female and male patients are of same age i.e there is no difference in terms of mean age between male and female patients

Testing sample proportion vs. population proportion(z – test for large samples)

Hypothesis Testing sample proportion vs. population proportion

```
library("xlsx")
##Import dataset into r
z <- read.xlsx("salesdata.xlsx",1,check.names=FALSE)
names(z) <- gsub(" ","", names(z))
#Hypothesis testing sample proportion with population proportion
#We would like to test the whether the proportion of customers
recommendation rate in the sample #salesdata dataset is 0.6
#H0 : p = 0.6
#H1 : p <> 0.6
mp1<- mean(z$Recommend)
sep1 <- sqrt(mp1*(1-mp1)/length(z$Recommend))
mz1 <- (mp1-0.6)/sep1
pz1 <- 2*pnorm(-abs(mz1))
cat("Mean Proportion of customers who recommended =",mp1,"\n")
cat("Standard deviation Proportion of customers who recommended
=",sep1,"\n")
cat("p value=",pz1,"\n")
```

Hypothesis Testing sample proportion vs. population proportion

Mean Proportion of customers who recommended = 0.5337079
Standard deviation Proportion of customers who recommended = 0.02643966
p value= 0.01216577

Inference: As the p value is less than 0.05(0.009) we reject the null hypothesis that the survival rate is not equal to 0.6 and accepts the alternative hypothesis H1

Testing difference between two sample proportions (z – test for large samples)

Hypothesis Testing difference between two proportions

```
library("xlsx")
##Import dataset into r
z <- read.xlsx("salesdata.xlsx",1,check.names=FALSE)
names(z) <- gsub(" ","", names(z))
#Hypothesis testing difference between two proportions
#We would like to test the whether the mean recommendation rate of
the male customers is #different from the female customers in the
sample salesdata dataset
#H0 : malerecom = femalerecom
#H1 : malerecom <> femalerecom
rf <- z$Recommend[gen == 2]
rm <- z$Recommend [gen == 1]
sf <- length(rf)
sm <- length(rm)
pcomb <- (sum(rf)+sum(rm))/(sf+sm)
s1<-(sum(rf)/sf) - (sum(rm)/sm)
s2<-pcomb*(1-pcomb)
s3<-(1/sf)+(1/sm)
zpdiff <- s1/sqrt(s2*s3)
pzdiff <- 2*pnorm(-abs(zpdiff))
```

Hypothesis Testing difference between two proportions output

Male customers Mean recommendation rating = 0.5166667
Female customers Mean recommendation rating = 0.5511364
p value= 0.5145207

Inference: As the p value is greater than 0.05(0.514) we accept the null hypothesis that the recommendation rate is equal among male and female customers

Student T-test

Student t test is similar to z-test and it will help to test whether the population mean is equal to a value or difference between two sample means when the sample size is small (i.e less than 30)

Hypothesis Testing t test for sample mean vs population mean and difference between two means

```
#Hypothesis testing sample mean vs. population mean
#We would like to test the mean age of the customers in the sample
salesdata dataset is 45
#H0 : μ = 45
#H1 : μ <> 45
#Hypothesis testing two means
#We would like to test the whether there is any significance difference
between the #mean age of male and female customers in sample
salesdata dataset
#H0 : meanagemale = meanagefemale
#H1 : meanagemale <> meanagefemale
t.test(agef,agem)
t.test(agef,)
```

t-test for testing sample mean vs. population mean output and difference between two population means

One Sample t-test
data: agef
t = 32.811, df = 175, p-value < 2.2e-16
alternative hypothesis: true mean is not equal to 0
95 percent confidence interval:
37.92507 42.77947
sample estimates:
mean of x
40.35227
Welch Two Sample t-test
data: agef and agem
t = -2.2766, df = 352.1, p-value = 0.02341
alternative hypothesis: true difference in means is not equal to 0
95 percent confidence interval:
-7.1613187 -0.5230247
sample estimates:
mean of x mean of y
40.35227 44.19444

Inference: As the p value is less than 0.05(0.0 and 0.02) we reject both the null hypothesis that the female age is 45 and female and male customers are of same age i.e there is no difference in terms of mean age between male and female customers

Chi-Square Test

Chi square test helps to carry out goodness s of fit test for the distribution of a particular variable and also helps to carry out independence test between two categorical variables. The independence test is referred as contingency test or classification test also.

Hypothesis Testing ChiSquare test for testing whether survival rate between male and female patients are same or survival and gender are independent

#Hypothesis testing – Whether gender and recommendation rating are independent of each other

#H0 : Recommendation rating and gender are independent

#H1: Recommendation rating and gender are not independent

we need to create two way table from the data set

table1 <- table(z$gender,z$Recommend)

chisq.test(table1)

t-test for testing sample mean vs. population mean output

table1

0 1

1 87 93

2 79 97

Pearson's Chi-squared test with Yates' continuity correction

data: table1

X-squared = 0.29764, df = 1, p-value = 0.5854

Inference: As the p value is greater than 0.05(0.5854) we accept both the null hypothesis that the recommendation rating and gender are independent or in other words the recommendation rating is same for both male and female customers

Shapiro Wilk normality Test

Shapiro normality test helps to test whether the variable follows normal distribution are not

Hypothesis Testing normality test - Shapiro Wilk normality test

#Hypothesis testing - Whether age follows normal distribution or nor

#H0 : Age follows normal distribution

#H1: Age does not follows normal distribution

shapiro.test(age)

Shapiro wilks normality test output

Shapiro-Wilk normality test

data: age

W = 0.95507, p-value = 5.786e-09

Inference: As the p value is less than 0.05(0.00005) we reject null hypothesis and conclude that the age does not follow normal distribution.

References

1. Tsagris, M., Alenazi, A., Verrou, K. M., & Pandis, N. (2020). Hypothesis testing for two population means: parametric or non-parametric test?. Journal of statistical computation and simulation, 90(2), 252-270.
2. Kääriäinen, M., Mikkonen, K., & Kyngäs, H. (2020).

Statistical Testing of a Theory. In The Application of Content Analysis in Nursing Science Research (pp. 95-103). Springer, Cham.

3. Chidlow, A., Greene, W., & Miller, S. R. (2020). What's in a p? A Commentary. In Research Methods in International Business (pp. 111-117). Palgrave Macmillan, Cham.
4. R Core Team (2018). R: A language and environment for statistical computing. R Foundation for Statistical Computing, Vienna, Austria. URL https://www.R-project.org/.
5. https://www.rstudio.com/
6. Casella, G., & Berger, R. L. (2002). Statistical inference (Vol. 2). Pacific Grove, CA: Duxbury.
7. Silvey, S. D. (2017). Statistical inference. Routledge.
8. Walker, H. M., & Lev, J. (1953). Statistical inference. Oxford And IBH Publishing Co.; Calcutta
9. Azzalini, A. (2017). Statistical inference based on the likelihood. Routledge.
10. William, L. T. (2000). Null hypothesis testing: problems, prevalence, and an alternative. J. Wildl. Manage, 64(4), 912-923.
11. Gardner, M. J., & Altman, D. G. (1986). Confidence intervals rather than P values: estimation rather than hypothesis testing. Br Med J (Clin Res Ed), 292(6522), 746-750.
12. Newey, W. K., & McFadden, D. (1994). Large sample estimation and hypothesis testing. Handbook of econometrics, 4, 2111-2245.
13. Mooi, E., Sarstedt, M., & Mooi-Reci, I. (2018). Hypothesis Testing & ANOVA. In Market Research (pp. 153-214). Springer, Singapore.
14. Ganeshpurkar, A., Maheshwari, R., Tekade, M., & Tekade, R. K. (2018). Concepts of Hypothesis Testing and Types of

Errors. In Dosage Form Design Parameters (pp. 257-280). Academic Press.

15. Haynes, W. (2013). Student's t-test. In Encyclopedia of Systems Biology (pp. 2023-2025). Springer, New York, NY.
16. Johnston, L. W. (1970). Student's t-Test. Journal of Quality Technology, 2(4), 243-245.
17. Welch, B. L. (1938). The significance of the difference between two means when the population variances are unequal. Biometrika, 29(3/4), 350-362.

Chapter 6 - Analysis of Variance

When we want to test difference between means of more than two groups, Analysis of Variance model can be used to test the same. In the sales data set if we want to test the difference in average quantity bought (response) among the income groups (factors) we can use the one way analysis variance (only one independent variable is used). Two way analysis variance models are used when we use two categorical variables such income group (3 categories) and gender (2 categories) and their effect on the average quantity bought. Here we can test the effect of interaction between two variables income groups and gender on mean quantity bought (3*2 combinations). That is whether is there any difference in mean quantity bought between group1,2,3 & female customers and group1,2,3 & male customers.

After conducting ANOVA, Post Hoc and multiple comparison tests can be used to find which of the means differ among them instead of conducting series of t-test to check whether which means differ among themselves and Post Hoc test will also reduce the type I error.

One Way Analysis of Variance (ANOVA)

one way analysis variance

```
library("xlsx")
##Import dataset into r
z <- read.xlsx("salesdata.xlsx",1,check.names=FALSE)
names(z) <- gsub(" ","", names(z))
#Hypothesis testing – Whether average quantity bought differ among the 3 income groups
#Ho: Average quantity bought is equal among the groups
#H1: Average quantity bought differ among the groups
#create factor variable from
gr<-z$incomerange
pr<-z$quantity
gr1<-factor(gr)
summary(lm(pr~gr1))
```

one way analysis variance output

```
Call:
lm(formula = pr ~ gr1)
Residuals:
Min 1Q Median 3Q Max
-1.925 -0.925 0.075 1.112 2.283
Coefficients:
Estimate Std. Error t value Pr(>|t|)
(Intercept) 2.92500 0.12236 23.904 <2e-16 ***
gr12 -0.20833 0.17305 -1.204 0.229
gr13 -0.03707 0.17453 -0.212 0.832
---

SIGNIF. CODES: 0 '***' 0.001 '**' 0.01 '*' 0.05 '.' 0.1 ' ' 1
Residual standard error: 1.34 on 353 degrees of freedom
Multiple R-squared: 0.004639, Adjusted R-squared: -0.0009999
F-statistic: 0.8227 on 2 and 353 DF, p-value: 0.4401
```

Inference: As the p value is greater than the 0.05 (0.4401) we accept the null hypothesis that there is no difference in average quantity bought among the three income groups.

TWO WAY ANALYSIS OF Variance (ANOVA)

Two way analysis variance

```
library("xlsx")
##Import dataset into r
z <- read.xlsx("clinical1.xlsx",1,check.names=FALSE)
names(z) <- gsub(" ","", names(z))
#Hypothesis testing – Whether average quantity bought differ among the 3 income groups and gender
#Ho: Average quantity bought equal among the 3 income groups and gender
#H1: Average quantity bought differ among the 3 income groups and gender
#create factor variable from
gr1<-factor(gr)
genf1<-factor(z$gender)
summary(lm(pr~gr1+genf1+gr1*genf1))
```

Two way analysis variance output

Call:
lm(formula = pr ~ gr1 + genf1 + gr1 * genf1)
Residuals:
Min 1Q Median 3Q Max
-2.00000 -0.93651 0.06349 1.11240 2.37705
Coefficients:
Estimate Std. Error t value Pr(>|t|)
(Intercept) 2.93651 0.16925 17.350 <2e-16 ***
gr12 -0.12295 0.24339 -0.505 0.614
gr13 -0.16065 0.24447 -0.657 0.512
genf12 -0.02423 0.24558 -0.099 0.921
gr12:genf12 -0.16638 0.34711 -0.479 0.632
gr13:genf12 0.24837 0.35006 0.709 0.478
—-

SIGNIF. CODES: 0 '***' 0.001 '**' 0.01 '*' 0.05 '.' 0.1 ' ' 1
Residual standard error: 1.343 on 350 degrees of freedom
Multiple R-squared: 0.008664, Adjusted R-squared: -0.005498
F-statistic: 0.6117 on 5 and 350 DF, p-value: 0.691

Inference: As the p value is greater than the 0.05 (0.691) we accept the null hypothesis that there is no difference in quantity bought with respect to gender and the groups. Here both the main and interaction effects are also not significant.

Post HOC Comparison Tests

Post HOC tests help us to know which treatments are differing when we obtain a significant p values from ANOVA. It helps us to control the type I error while comparing the two means. Instead of conducting one Post HOC test if we conduct series of t test to check which of the two means it will increase the Type I error multiple times the number of tests conducted.

The following are normally used Post Hoc Comparison Tests

Fisher's LSD test

Fishers's Least Significant Difference test is most liberal among the Post Hoc tests

Tukey's HSD test

Tukey's Honestly Significant Difference test is conservative test in controlling the Type I error and useful when we have more number of comparisons to test

Dunn's Test

Dunn's Test is a conservative test and reduces power when the number of comparison increases

Scheffe test

Scheffe test is most conservative of all tests. It helps us to make simple and complex comparisons.

Newmankeuls test

It is a step down procedure and initially difference between the largest mean and smallest mean is tested if it found significant then the smallest mean is tested with the next largest mean. This procedure is repeated until a non-significant result is obtained

Let us consider the following data set cropyield.xlsx which can be used to explain the Post HOC Tests which can be downloaded from author's website www.ijsmi.com/book.php[1]. The following is the structure of the data set

- Plotnumber
- Crop
- Fertilizer
- Yield

One Way ANOVA and Post HOC comparison test

```
library("xlsx")
library(DescTools)
z <- read.xlsx("cropyield.xlsx",1,check.names=FALSE)
summary(lm(z$Yield ~ as.factor(z$Fertilizer)))
PostHocTest(aov(z$Yield ~ as.factor(z$Fertilizer), data = z), method
= "lsd")
```

1. http://www.ijsmi.com/book.php

One Way ANOVA output

Call:

lm(formula = z$Yield ~ as.factor(z$Fertilizer))

Residuals:

Min 1Q Median 3Q Max

-497.69 -57.69 -2.03 95.14 282.86

Coefficients:

Estimate Std. Error t value Pr(>|t|)

(Intercept) 856.36 38.63 22.170 < 2e-16 ***

as.factor(z$Fertilizer)2 -389.22 61.94 -6.284 2.33e-07 ***

as.factor(z$Fertilizer)3 -441.82 54.63 -8.088 8.74e-10 ***

as.factor(z$Fertilizer)4 31.33 52.48 0.597 0.554

SIGNIF. CODES: 0 '***' 0.001 '**' 0.01 '*' 0.05 '.' 0.1 ' ' 1

Residual standard error: 128.1 on 38 degrees of freedom

Multiple R-squared: 0.7617, Adjusted R-squared: 0.7429

F-statistic: 40.49 on 3 and 38 DF, p-value: 6.438e-12

Inference: The crop yield differed with respect to the different fertilizers as p value was less than 0.05 (p-value: 6.438e-12)

When we apply LSD Post Hoc Test the following results were obtained

Post HOC comparison test output

Posthoc multiple comparisons of means : Fisher LSD
95% family-wise confidence level
$`as.factor(z$Fertilizer)`
diff lwr.ci upr.ci pval
2-1 -389.22078 -514.61640 -263.82516 2.3e-07 ***
3-1 -441.81818 -552.40672 -331.22964 8.7e-10 ***
4-1 31.32867 -74.92136 137.57870 0.5541
3-2 -52.59740 -177.99302 72.79821 0.4011
4-2 420.54945 298.96284 542.13606 2.4e-08 ***
4-3 473.14685 366.89682 579.39688 5.6e-11 ***
—-

SIGNIF. CODES: 0 '***' 0.001 '**' 0.01 '*' 0.05 '.' 0.1 ' ' 1

Inference: From the above it can be found that

Effect of fertilizer 4 on the crop yield is different from the fertilizers 2 and 3 not different from 1.

Effect of fertilizer 1 on the crop yield is different from the fertilizers 2 and 3.

Effect of fertilizer 2 on the crop yield is not different from the fertilizers 3

Readers are encouraged to test the remaining Post Hoc tests by changing the method in PostHocTest function

PostHocTest(aov(z$Yield ~ as.factor(z$Fertilizer), data = z), method = "xxx").

References

1. Stoker, P., Tian, G., & Kim, J. Y. (2020). 11 Analysis of Variance (ANOVA). Basic Quantitative Research Methods for Urban Planners.
2. Kallner, A., & Theodorsson, E. (2020). An experimental study of methods for the analysis of variance components in the inference of laboratory information. Scandinavian Journal of Clinical and Laboratory Investigation, 80(1), 73-80.
3. R Core Team (2018). R: A language and environment for statistical computing. R Foundation for Statistical Computing, Vienna, Austria. URL https://www.R-project.org/.
4. https://www.rstudio.com/
5. Iversen, G. R., Wildt, A. R., Norpoth, H., & Norpoth, H. P. (1987). Analysis of variance (No. 1). Sage.
6. Penny, W., & Henson, R. (2006). Analysis of variance. Statistical parametric mapping: The analysis of functional brain images, 166-177.
7. Barr, A. J., Goodnight, J. H., Sail, J. P., & Helwig, J. T. (1976). Analysis of variance. SAS.
8. Kass, R. E., Eden, U. T., & Brown, E. N. (2014). Analysis of variance. In *Analysis of Neural Data* (pp. 361-389). Springer, New York, NY.
9. Ruxton, G. D., & Beauchamp, G. (2008). Time for some a priori thinking about post hoc testing. Behavioral Ecology, 19(3), 690-693.
10. Pohlert, T. (2014). The pairwise multiple comparison of mean ranks package (PMCMR). R package, 27.
11. Cardinal, R. N., & Aitken, M. R. (2013). ANOVA for the behavioral sciences researcher. Psychology Press.
12. Hilton, A., & Armstrong, R. (2006). Statnote 6: post-hoc ANOVA tests.
13. Kim, H. Y. (2015). Statistical notes for clinical researchers:

post-hoc multiple comparisons. Restorative dentistry & endodontics, 40(2), 172-176.

Chapter 7 - Correlation and Regression

Correlation

Correlation helps to find the relationship exists between two or more variables and it plays an important part in the clinical research. The correlation is measured through Pearson correlation coefficient which takes value between 1(positive correction) and -1 (negative correlation). Correlation coefficient value 0 means there is no correlation between the two variables and if it is 1 or -1 it indicates the perfect correlation positive and negative respectively.

Correlation does not mean that there is a cause and effect relationship between the variables. The correlation coefficient can be tested to know whether it is differ significantly from 0 through correlation test

Correlation between two variables

#We would like to know correlation coefficient between the age and quantity bought and the correlation coefficient is significant or not

H0: There is no correlation between age and quantity bought

H1: There is a correlation between age and quantity bought

```
cor(age,z$quantity)
cor.test(age,z$quantity)
```

Correlation between two variables output

```
cor(age,z$quantity)
[1] -0.008734007
> cor.test(age,z$quantity)
Pearson's product-moment correlation
data: age and z$quantity
t = -0.16434, df = 354, p-value = 0.8696
alternative hypothesis: true correlation is not equal to 0
95 percent confidence interval:
-0.11257340 0.09529409
sample estimates:
cor
-0.008734007
```

Here the correlation coefficient value indicates that there is mild negative correlation between the variable age and quantity bought. But the p value is less than 0.05 we reject null hypothesis that there is no significant correlation between the variables.

Regression

Regression helps to model the relationship between the dependent variables and the independent variables. The regression equation tries to minimize the errors between the predicted values and the actual values.

There are different types of regression models are available based on the nature of the data such as simple linear regression, multiple linear regression, non-linear regression, logistic regression, Quantile regression

Simple Linear Regression

Simple linear regression helps to model the relationship between one dependent variable and one independent variable. It helps us to predict the value of the dependent variable based on the independent variable. The error term is defined as the difference between predicted and the actual value.

The following are assumptions are considered during the modeling process .

- The relation between the variables are linear in nature
- The variables follow normal distribution
- The error terms are equal for any value of the independent variable

The simple linear regression equation is defined by the following equation

$Y = a+bx+e$

- Y = Dependent variable
- a= constant or intercept
- b= slope
- x = independent variable
- e = residual error term

The objective of the regression equation to estimate the parameters a and b by minimizing residual error

Simple linear regression

```
#We would like to model the relationship between the Quantity bought and age variables and predict the quantity bought values from the regression equation
summary(lm(quantity~age))
```

Simple linear regression output

Call:
lm(formula = z$quantity ~ age)
Residuals:
Min 1Q Median 3Q Max
-1.8626 -0.8548 0.1469 1.1564 2.1754
Coefficients:
Estimate Std. Error t value Pr(>|t|)
(Intercept) 2.8736191 0.2011524 14.286 <2e-16 ***
age -0.0007311 0.0044489 -0.164 0.87
—-

SIGNIF. CODES: 0 ‘***’ 0.001 ‘**’ 0.01 ‘*’ 0.05 ‘.’ 0.1 ‘ ’ 1
Residual standard error: 1.342 on 354 degrees of freedom
Multiple R-squared: 7.628e-05, Adjusted R-squared: -0.002748
F-statistic: 0.02701 on 1 and 354 DF, p-value: 0.8696

Here the regression equation is defined by

Quantity = 2.873+ (-0.0007)*age

The following needs to be observed from the regression equation output

The adjusted r-squared value

It is important value to judge the accuracy of the model fit. If the value of adjusted r square is above 0.8 is very good, 0.6-0.8 is good, and 0.3-0.6 is average and below 0.3 is below average. In the above model it is 0.002 which is poor.

P value of the parameters to be significant for a good model

Here the intercept is significant (<0.05) and age is not significant

Multiple Linear Regression

Multiple linear regression is similar to simple linear regression except the number of independent variables are more than one.

Multiple linear regression

#We would like to model the relationship between the LDL and age, gender, income range , price, rating, Recommend variables and predict the quantity values from the regression equation. When we are using the categorical variables, dummy variable or contrast to be used in the place of categorical variables. This is can be done using the factor(variable name) in R

```
lm(formula = z$quantity ~ age + factor(z$gender) +
factor(z$incomerange) +

z$price + z$rating + z$Recommend)
```

Multiple linear regression output

Residuals:

Min 1Q Median 3Q Max

-2.09695 -0.92694 0.06957 1.12531 2.40732

Coefficients:

Estimate Std. Error t value Pr(>|t|)

(Intercept) 3.1742588 0.3306167 9.601 <2e-16 ***

age -0.0009524 0.0045490 -0.209 0.834

factor(z$gender)2 -0.0063868 0.1442244 -0.044 0.965

factor(z$incomerange)2 -0.2077848 0.1741114 -1.193 0.234

factor(z$incomerange)3 -0.0399104 0.1763548 -0.226 0.821

z$price -0.0008461 0.0009898 -0.855 0.393

z$rating -0.0451404 0.0501078 -0.901 0.368

z$Recommend 0.0528867 0.1442804 0.367 0.714

—-

SIGNIF. CODES: 0 '***' 0.001 '**' 0.01 '*' 0.05 '.' 0.1 ' ' 1

Residual standard error: 1.347 on 348 degrees of freedom

Multiple R-squared: 0.009158, Adjusted R-squared: -0.01077

F-statistic: 0.4595 on 7 and 348 DF, p-value: 0.8635

Here the regression equation is defined by

3.1742588+(-0.0009524) * age +(-0.0063868) * gender + (-0.2077848)* incomerange2 + (-0.0399104)*incomerange3 + (-0.0008461)*price+ (-0.0451404) * rating + 0.0528867 * Recommend

The following needs to be observed from the regression equation output

The adjusted r-squared value

It is important value to judge the accuracy of the model fit. If the value of adjusted r square is above 0.8 is very good, 0.6-0.8 is good, and 0.3-0.6 is average and below 0.3 is below average. In the above model it is -0.0107 which is poor.

P value of the parameters to be significant for a good model

Here the intercept is significant (<0.05) and all the other variables are not significant

Logistic regression

Logistics regression is used when the dependent variable is binary categorical variable which takes either of the two values. In the logistics regression, the odds of binary categorical dependent variable (response) taking an either of the particular value is depending on the independent variables (predictor). Interpretation of the logistic regression coefficients is not straightforward as the coefficients in the model give us the change in the log odds of the dependent variable for a one unit increase in the independent variable.

Logistics Regression

#We would like to model the relationship between the Recommendati
and age, gender, income range , price variables. When we are u
categorical variables, dummy variable or contrast to be used in the
categorical variables. This is can be done using the factor(variable nam

```
logist <- glm(z$Recommend
z$quantity+age+factor(z$gender)+factor(z$incomerange)+z$price+
data = z, family = "binomial")

summary(logist)

exp(coef(logist))
```

Logistics regression output

Call:

glm(formula = z$Recommend ~ z$quantity + age + factor(z$gender) +

factor(z$incomerange) + z$price + z$rating, family = "binomial",

data = z)

Deviance Residuals:

Min 1Q Median 3Q Max

-1.5095 -1.2115 0.9474 1.1103 1.3622

Coefficients:

Estimate Std. Error z value Pr(>|z|)

(Intercept) 0.363197 0.542237 0.670 0.5030

z$quantity 0.029904 0.080484 0.372 0.7102

age -0.013321 0.006813 -1.955 0.0506 .

factor(z$gender)2 0.084253 0.216297 0.390 0.6969

factor(z$incomerange)2 -0.047731 0.261099 -0.183 0.8549

factor(z$incomerange)3 0.222270 0.265112 0.838 0.4018

```
z$price 0.001046 0.001486 0.704 0.4815

z$rating 0.009630 0.075255 0.128 0.8982

—-

Signif. codes: 0 '***' 0.001 '**' 0.01 '*' 0.05 '.' 0.1 ' ' 1

(Dispersion parameter for binomial family taken to be 1)

Null deviance: 491.90 on 355 degrees of freedom

Residual deviance: 485.93 on 348 degrees of freedom

AIC: 501.93

Number of Fisher Scoring iterations: 4

> exp(coef(logist))

(Intercept) z$quantity age

1.4379195 1.0303554 0.9867676

factor(z$gender)2 factor(z$incomerange)2 factor(z$incomerange)3

1.0879038 0.9533901 1.2489081

z$price z$rating

1.0010468 1.0096770
```

Here the regression equation is defined by

1.43+1.0303*quantity+0.98*ae+1.087*gender2+ 0.953*incomerange2 + 1.24*incomerange3+ 1.001*price + 1.009*rating

Here the raw regression coefficients represent change in log odds of recommended (vs not recommended) with respect to one unit change in independent variable. For example one unit change in age, log odds of the recommendation will change by 0.029904

When we take the exponentiation of the coefficients it gives the odds ratio of survival with respect to one unit change in the independent variables. For example one unit change in age, odds of survival will increases by 0.9867676.

The following needs to be observed from the regression equation output

P value of the parameters to be significant for a good model

- Here all variables are not significant (p>0.0.5)

Non Linear Regression

Non Linear regression is used when the relationship between the predictor and the response variable is nonlinear. The model requires the nonlinear relationship to be defined before with the starting value for the parameters.

Non Linear Regression

#We would like to model the relationship between the Quantity bought and age using the qubic form. When we are using the categorical variables, dummy variable or contrast to be used in the place of categorical variables. This is can be done using the factor(variable name) in R

```
nlr <- nls(z$quantity ~ b3*age^3+b2*age^2+b1,start = list(b1 = 3,b2 = 7,b3=8),data=z)

summary(nlr)
```

Non Linear Regression output

Formula: z$quantity ~ b3 * age^3 + b2 * age^2 + b1

Parameters:

Estimate Std. Error t value Pr(>|t|)

b1 2.827e+00 2.298e-01 12.302 <2e-16 ***

b2 5.822e-05 3.771e-04 0.154 0.877

b3 -9.551e-07 5.299e-06 -0.180 0.857

—-

Signif. codes: 0 '***' 0.001 '**' 0.01 '*' 0.05 '.' 0.1 ' ' 1

Residual standard error: 1.343 on 353 degrees of freedom

Number of iterations to convergence: 1

Achieved convergence tolerance: 4.293e-08

Here the intercept is significant and other coefficients are not significant

Quantile Regression

Quantile regression is used when the normality and equal variance assumption is violated. We need to use quantreg package for carrying out the quantile regression.

Quantile Regression

#We would like to model the relationship between the Quantity bought and age using quantile regression.

```
quantreg1 <- rq(z$quantity ~ age, tau = .5, data = z)

quantreg1
```

Quantile Regression output

```
quantreg1

Call:

rq(formula = z$quantity ~ age, tau = 0.5, data = z)

Coefficients:

(Intercept) age

3.000000e+00 4.163336e-18

Degrees of freedom: 356 total; 354 residual
```

HERE THE QUANTILE REGRESSION equation is represented as

Quantity = 3.0+(4.163336e-18) *age

We can also plot the regression lines of linear and quantile regression as below

Quantile Regression

plot(z$quantity ~ age, data=z , main = "Regression")

abline(lm (z$quantity ~ age, data = z), col = "Brown")

abline(rq(z$quantity ~ age, data = z), col = "Yellow")

legend("topright", legend = c("lm", "rq"), col = c("Brown", "Yellow"))

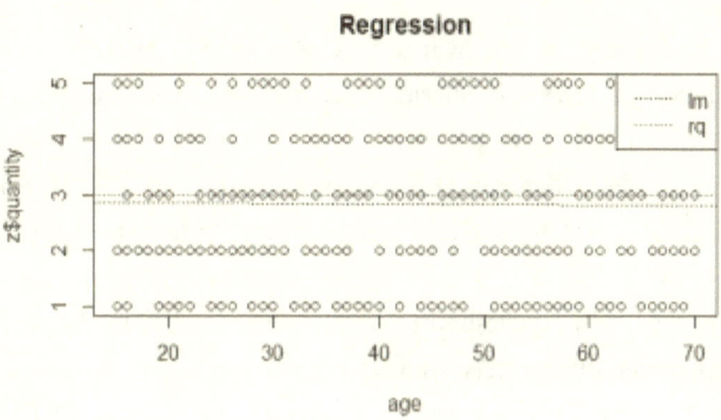

References

1. R Core Team (2018). R: A language and environment for statistical computing. R Foundation for Statistical Computing, Vienna, Austria. URL https://www.R-project.org/.
2. https://www.rstudio.com/
3. Crawford, S. L. (2006). Correlation and regression. Circulation, 114(19), 2083-2088.

4. Benesty, J., Chen, J., Huang, Y., & Cohen, I. (2009). Pearson correlation coefficient. In Noise reduction in speech processing (pp. 1-4). Springer, Berlin, Heidelberg.
5. Sedgwick, P. (2012). Pearson's correlation coefficient. BMJ: British Medical Journal (Online), 345.
6. Wang, J. (2013). Pearson correlation coefficient. In Encyclopedia of systems biology (pp. 1671-1671). Springer, New York, NY.
7. Nahler, G. (2009). Pearson correlation coefficient. In Dictionary of Pharmaceutical Medicine (pp. 132-132). Springer, Vienna.
8. Hall, G. (2015). Pearson's correlation coefficient. other words, 1(9).
9. Lee Rodgers, J., & Nicewander, W. A. (1988). Thirteen ways to look at the correlation coefficient. The American Statistician, 42(1), 59-66.
10. Zou, K. H., Tuncali, K., & Silverman, S. G. (2003). Correlation and simple linear regression. Radiology, 227(3), 617-628.
11. Kutner, M. H., Nachtsheim, C., & Neter, J. (2004). Applied linear regression models. McGraw-Hill/Irwin.
12. Montgomery, D. C., Peck, E. A., & Vining, G. G. (2012). Introduction to linear regression analysis (Vol. 821). John Wiley & Sons.
13. Neter, J., Wasserman, W., & Kutner, M. H. (1989). Applied linear regression models.
14. Pedhazur, E. J., & Kerlinger, F. N. (1973). Multiple regression in behavioral research. New York: Holt, Rinehart and Winston.
15. Aiken, L. S., West, S. G., & Reno, R. R. (1991). Multiple regression: Testing and interpreting interactions. Sage.
16. Allison, P. D. (1999). Multiple regression: A primer. Pine

Forge Press.
17. Sperandei, S. (2014). Understanding logistic regression analysis. Biochemia medica: Biochemia medica, 24(1), 12-18.
18. Morrissey, M. B., & Ruxton, G. D. (2018). Multiple regressions: the meaning of multiple regression and the non-problem of collinearity. Philosophy, Theory and Practice in Biology.
19. Peng, C. Y. J., Lee, K. L., & Ingersoll, G. M. (2002). An introduction to logistic regression analysis and reporting. The journal of educational research, 96(1), 3-14.
20. Koenker, R., & Hallock, K. F. (2001). Quantile regression. Journal of economic perspectives, 15(4), 143-156.
21. Yu, K., Lu, Z., & Stander, J. (2003). Quantile regression: applications and current research areas. Journal of the Royal Statistical Society: Series D (The Statistician), 52(3), 331-350.

Chapter 7 - Non Parametric Tests

Nonparametric tests do not assume any particular distribution whereas the parametric tests assume some specific distribution for the given dataset. Non Parametric tests are also called as distribution free tests.

Mann-Whitney U test

Mann-Whitney U test helps us to test whether two sample means are equal. This test is equal to parametric test – t test for testing two sample means.

Let us consider the following data set salesdata for carrying out Mann Whitney u test to test whether there is any difference in recommended rating among male and female customers

Mann-Whitney U test

```
library("xlsx")
wd = "C:\\test"
setwd(wd)
z <- read.xlsx("salesdata.xlsx",1,check.names=FALSE)
names(z) <- gsub(" ","", names(z))
wilcox.test(z$Recommend~z$gender,z)
```

Mann-Whitney U test output

```
Wilcoxon rank sum test with continuity correction
data: z$Recommend by z$gender
W = 15294, p-value = 0.5155
alternative hypothesis: true location shift is not equal to
```

Inference: From the above results, we can infer that recommended rating is not different among male and female patients as p value is greater than 0.05 (p- 0.5155)

Wilcoxon Signed Rank Test

Wilcoxon Signed rank test helps us to test two matched samples means are equal or not. It is equivalent of parametric test - paired t-test

Let us consider the following data set exercise.xlsx on pre and post exercise with the following variables. The data set can be downloaded from the author's website (www.ijsmi.com/dataset.php[1])

Id

Age

Gender

Exercisestatus(coded as 1 = carried out exercise , 2 = not done exercise)

Preweight

Postweight

Using the Wilcoxin signed rank test we can test the hypothesis that whether the exercise has an effect in reducing the body weight

Wilcoxon Signed Rank Test

```
library("xlsx")
wd = "C:\\test"
setwd(wd)
z <- read.xlsx("exercise.xlsx",1,check.names=FALSE)
names(z) <- gsub(" ","", names(z))
preweight<- z$preweight
postweight<- z$postweight
```

1. http://www.ijsmi.com/dataset.php

Wilcoxon Signed Rank Test output

```
Wilcoxon signed rank test with continuity correction
data: preweight and postweight
V = 828, p-value = 0.001698
alternative hypothesis: true location shift is not equal to 0
```

Inference: From the above results, we can infer that exercise has an effect on reducing the body weight as p value is less than 0.05 (p-0.001698)

Kruskal Wallis Test

Kruskal Wallis test helps us to test whether more than two sample means are equal or not. It is equivalent of parametric test – One way ANOVA test.

Let us consider the following data set salesdata.xlsx for carrying out Kruskal Wallis Test to test whether there is any difference in quantity bought among different income groups

Kruskal Wallis Test

```
library("xlsx")
wd = "C:\\test"
setwd(wd)
z <- read.xlsx("exercise.xlsx",1,check.names=FALSE)
names(z) <- gsub(" ","", names(z))
kruskal.test(z$quantity~z$incomerange,z)
```

Kruskal Wallis Test output

```
Kruskal-Wallis rank sum test
data: z$quantity by z$incomerange
Kruskal-Wallis chi-squared = 1.1271, df = 2, p-value = 0.5692
```

INFERENCE: FROM THE above results, we can infer that quantiy bought is not different among different income groups as p value is greater than 0.05 (p- 05692)

References

1. le Cessie, S., Goeman, J. J., & Dekkers, O. M. (2020). Who is afraid of non-normal data? Choosing between parametric and non-parametric tests. European Journal of Endocrinology, 182(2), E1-E3.
2. Sant'Anna, P. H. (2020). Nonparametric tests for treatment effect heterogeneity with duration outcomes. Journal of Business & Economic Statistics, (just-accepted), 1-56.
3. Siegel, S. (1957). Nonparametric statistics. The American Statistician, 11(3), 13-19.
4. Varian, H. R. (1983). Non-parametric tests of consumer behaviour. The review of economic studies, 50(1), 99-110.
5. Deheuvels, P. (1980). Non parametric tests of independence. In Statistique non paramétrique asymptotique (pp. 95-107). Springer, Berlin, Heidelberg.
6. McKight, P. E., & Najab, J. (2010). Kruskal-wallis test. The corsini encyclopedia of psychology, 1-1.
7. Nachar, N. (2008). The Mann-Whitney U: A test for assessing whether two independent samples come from the same distribution. Tutorials in quantitative Methods for Psychology, 4(1), 13-20.
8. Woolson, R. F. (2007). Wilcoxon signed-rank test. Wiley encyclopedia of clinical trials, 1-3.
9. Anderson, N. H. (1961). Scales and statistics: parametric and nonparametric. Psychological bulletin, 58(4), 305.

Chapter 8 - Advanced Statistical Models

This chapter provides overview of advanced statistical models such as factor analysis, cluster analysis, discriminant analysis and survival analysis.

Factor Analysis

Factor analysis is used to reduce the dimensionality of the dataset under study by combining the variables into underlying factors. It also helps us to study the interrelationship between the variables. The factor analysis is carried out in two stages factor extraction and factor rotation. The factor extraction can be carried out normally using Principal Component Analysis method which aims to reduce the number of variables in the dataset.

Eigenvalues calculated during the PCA will help us to obtain the amount variance explained by each principal component. The Principal Components whose eigenvalues are above 1 are retained in the analysis.

The scree plot which is obtained when the eigenvalues are plotted against the principal components also helps to decide the number of factors or principal component to be retained in the analysis.

Factor rotation is carried out by different methods such as orthogonal rotation and oblique rotations which helps us to refine the factors extracted through PCA.

We will use hypothetical data set factor1.xlsx excel file which has 10 variables with numerical values.

Factor Analysis

```
f <- read.xlsx("factor1.xlsx",1,check.names=FALSE)

principal <- princomp(f, cor=TRUE)

summary(principal)

loadings(principal)

factors <- factanal(f, 3, rotation="varimax")

factors

screeplot(principal, npcs = min(10, length(principal)),

type = c("barplot", "lines"),

))
```

Factor analysis- output

> principal <- princomp(f, cor=TRUE)

> summary(principal)

Importance of components:

Comp.1 Comp.2 Comp.3 Comp.4 Comp.5

Standard deviation 1.3440523 1.2364184 1.1673694 1.1093886 1.0386250

Proportion of Variance 0.1806477 0.1528730 0.1362751 0.1230743 0.1078742

Cumulative Proportion 0.1806477 0.3335207 0.4697958 0.5928701 0.7007443

Comp.6 Comp.7 Comp.8 Comp.9 Comp.10

Standard deviation 0.97149685 0.82340043 0.74685236 0.67125915 0.60198429

Proportion of Variance 0.09438061 0.06779883 0.05577884 0.04505888 0.03623851

Cumulative Proportion 0.79512494 0.86292376 0.91870261 0.96376149 1.00000000

> loadings(principal)

Loadings:

Comp.1 Comp.2 Comp.3 Comp.4 Comp.5 Comp.6 Comp.7 Comp.8 Comp.9 Comp.10

v1 0.469 0.471 0.138 -0.234 0.244 0.392 -0.517

v2 0.156 -0.182 -0.556 0.234 -0.311 -0.145 0.616 -0.134 -0.175 -0.182

v3 0.293 -0.128 -0.166 -0.192 0.848 0.227 0.217

v4 -0.370 -0.140 0.387 -0.108 0.357 0.241 0.621 -0.151 -0.251 0.157

v5 0.336 -0.452 0.284 0.390 -0.377 0.500 0.216

v6 -0.406 -0.332 0.325 0.393 0.624 0.254

v7 0.132 0.764 0.322 -0.234 -0.478

v8 -0.449 -0.509 -0.125 -0.225 -0.168 -0.652

v9 -0.542 0.271 -0.501 0.418 0.441

v10 -0.207 0.415 0.288 0.348 -0.424 0.123 0.241 -0.138 0.556

Comp.1 Comp.2 Comp.3 Comp.4 Comp.5 Comp.6 Comp.7 Comp.8 Comp.9

SS loadings 1.0 1.0 1.0 1.0 1.0 1.0 1.0 1.0 1.0

Proportion Var 0.1 0.1 0.1 0.1 0.1 0.1 0.1 0.1 0.1

Cumulative Var 0.1 0.2 0.3 0.4 0.5 0.6 0.7 0.8 0.9

Comp.10

SS loadings 1.0

Proportion Var 0.1

Cumulative Var 1.0

> factors <- factanal(f, 3, rotation="varimax")

> factors

Call:

factanal(x = f, factors = 3, rotation = "varimax")

Uniquenesses:

v1 v2 v3 v4 v5 v6 v7 v8 v9 v10

0.092 0.826 0.916 0.846 0.650 0.882 0.987 0.005 0.768 0.775

Loadings:

Factor1 Factor2 Factor3

v1 0.949

v2 -0.101 0.403

v3 0.279

v4 0.253 -0.295

v5 0.323 0.209 0.450

v6 -0.332

v7 0.113

v8 -0.327 0.935 -0.116

v9 0.125 0.460

v10 -0.128 -0.452

Factor1 Factor2 Factor3

SS loadings 1.270 1.222 0.761

Proportion Var 0.127 0.122 0.076

Cumulative Var 0.127 0.249 0.325

Test of the hypothesis that 3 factors are sufficient.

The chi square statistic is 13.25 on 18 degrees of freedom.

The p-value is 0.777

Please refer authors book "Application of statistical tools in biomedical domain –an overview with help of software" for more details on factor analysis

Cluster Analysis

Cluster analysis helps us to identify natural clusters present in the dataset under consideration. The members in the same cluster will have similar properties wherein the members of between clusters will have different properties. There are two main types of clustering methods are available agglomerative clustering and divisive clustering.

Agglomerative clustering starts with single member per cluster initially and continue to merge with other cluster which closer to the given cluster and the same will continue till any more clusters can be merged. In the agglomerative clustering, the distance between the clusters serves as a medium to merge the two clusters which are closer and also based on the variance within the clusters. The distance between clusters is either measured based on the minimum, maximum and average distance.

Cluster Analysis – Agglomerative method – k means cluster

```
# Install rattle package and use k means divisive cluster methods
library('rattle')
kmean<- kmeans(f,3)
kmean$cluster
library(cluster)
clusplot(f, kmean$cluster)
```

Cluster Analysis – Agglomerative method – k means cluster output

Cluster method : complete
Distance : euclidean
Number of objects: 47

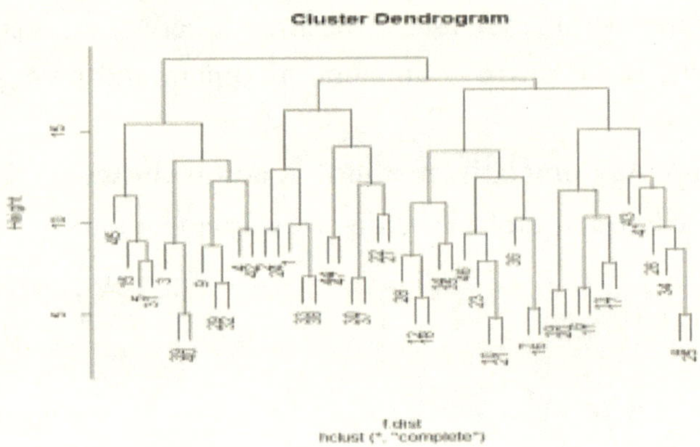

DIVISIVE CLUSTERING on the other hand, starts with the a single cluster comprising of all members of the dataset initially and starts dividing into different clusters while ensuring the members within the clusters are close and members between clusters are different and each members distance from the cluster centroid is less. In this method, cluster membership is reassigned till the distance from the cluster centroid is less and continued till the reassignment is not possible. In the divisive clustering method the number of clusters needs to be specified before the clustering process starts

Cluster Analysis – divisive method – k means cluster

```
# Install rattle package and use k means divisive cluster methods
library('rattle')
kmean<- kmeans(f,3)
kmean$cluster
library(cluster)
clusplot(f, kmean$cluster)
```

Cluster Analysis – divisive method – k means cluster output

K-means clustering with 3 clusters of sizes 17, 15, 15

Cluster means:

v1 v2 v3 v4 v5 v6 v7 v8 v9

1 6.411765 6.352941 6.176471 5.117647 6.823529 5.176471 5.529412 6.647059 8.176471

2 2.533333 5.066667 5.600000 6.866667 4.800000 7.533333 6.533333 7.400000 4.266667

3 6.466667 5.933333 6.933333 4.600000 5.000000 4.733333 5.400000 2.200000 2.800000

v10

1 4.058824

2 6.266667

3 6.266667

Clustering vector:

1 2 3 4 5 6 7 8 9 10 11 12 13 14 15 16 17 18 19 20 21 22 23 24 25 26 27

3 3 2 2 1 2 3 1 2 3 2 2 3 2 1 3 2 2 2 2 3 1 3 3 1 1 1

28 29 30 31 32 33 34 35 36 37 38 39 40 41 42 43 44 45 46 47

2 1 1 1 1 3 1 2 3 3 3 1 1 1 2 1 2 1 3 3

Within cluster sum of squares by cluster:

[1] 1122.7059 903.3333 797.4667

(between_SS / total_SS = 23.9 %)

Available components:

[1] "cluster" "centers" "totss" "withinss" "tot.withinss"

[6] "betweenss" "size" "iter" "ifault"

> kmean$cluster

1 2 3 4 5 6 7 8 9 10 11 12 13 14 15 16 17 18 19 20 21 22 23 24 25 26 27

3 3 2 2 1 2 3 1 2 3 2 2 3 2 1 3 2 2 2 2 3 1 3 3 1 1 1

28 29 30 31 32 33 34 35 36 37 38 39 40 41 42 43 44 45 46 47

2 1 1 1 1 3 1 2 3 3 3 1 1 1 2 1 2 1 3 3

> library(cluster)

Cluster Analysis – divisive method – k means cluster output

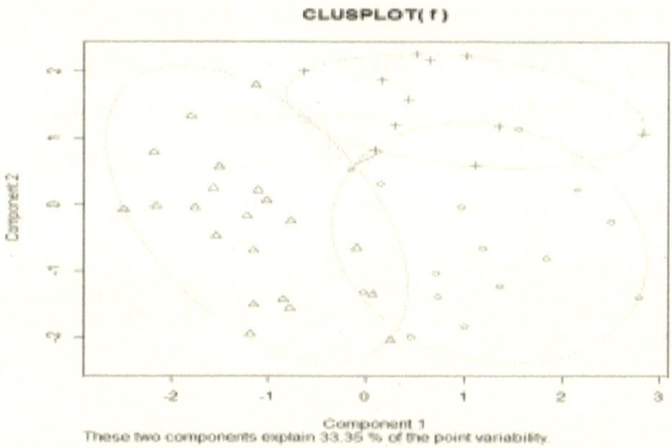

Discriminant Analysis

Discriminant analysis helps us to predict the group membership of the new cases using the discriminant function which is a linear combination of predictor variables created from the dataset having known group membership. The training dataset is used to create the discriminant function which is used to divide the dataset into groups and test dataset is used during the prediction of group membership of the new dataset.

Discriminant Analysis

Install MASS package to perform discriminant analysis. Split the dataset factor1 into two train and test dataset. The train dataset is used to create linear discriminant function and test dataset is used to predict the group membership of test data.

```
library(MASS)

data = f

sample1 = sample(1:nrow(f), size=0.25*nrow(f))

test = data[sample1,]

train = data[-sample1,]

mlda <- lda(v11~.,data = train)

predictions <- predict(mlda,test)
```

Discriminant Analysis output

```
> library(MASS)

> data = f

> indexes = sample(1:nrow(f), size=0.25*nrow(f))

> test = data[indexes,]

> train = data[-indexes,]

> mlda <- lda(v11~.,data = train)

> predictions <- predict(mlda,test)

> mlda

Call:

lda(v11 ~ ., data = train)

Prior probabilities of groups:

1 2 3

0.1388889 0.5833333 0.2777778

Group means:

v1 v2 v3 v4 v5 v6 v7 v8 v9 v10
```

1 5.600000 5.600000 7.600000 4.0 7.00000 4.200000 5.00000 5.0 6.000000 3.200000

2 4.952381 6.428571 5.952381 5.0 5.52381 6.095238 6.47619 5.0 5.285714 5.952381

3 6.200000 5.500000 5.400000 7.1 5.70000 6.500000 5.50000 6.9 5.500000 6.400000

Coefficients of linear discriminants:

LD1 LD2

v1 0.253466477 0.14508454

v2 0.127058378 -0.12838499

v3 -0.074757253 0.11599765

v4 0.194081026 0.03555130

v5 -0.062988872 0.05578483

v6 0.102451234 -0.07178563

v7 0.006307868 -0.12941158

v8 0.211171652 0.16901417

v9 -0.033242626 -0.05573464

v10 0.193818122 -0.11539288

Proportion of trace:

LD1 LD2

0.7541 0.2459

> predictions

$class

[1] 3 2 2 2 1 2 1 2 2 2 2

Levels: 1 2 3

$posterior

1 2 3

15 0.01401637 0.3857245 0.600259119

20 0.31052498 0.5774898 0.111985215

28 0.13180324 0.7536139 0.114582845

46 0.23036048 0.6784816 0.091157915

23 0.55127963 0.4046478 0.044072572

29 0.03856265 0.7948662 0.166571195

47 0.69226403 0.3041630 0.003572942

14 0.17634394 0.6475574 0.176098662

13 0.06056098 0.8754091 0.064029925

34 0.04938745 0.5489317 0.401680881

1 0.31415109 0.4572253 0.228623613

$x

LD1 LD2

15 1.1913713 0.42636364

20 -0.8528351 0.83015675

28 -0.5943275 0.02904277

46 -0.8898085 0.34996789

23 -1.4824606 1.05014385

29 0.0205203 -0.54478304

47 -2.8494175 0.08483594

14 -0.4369225 0.62133541

13 -0.6644091 -0.93478729

34 0.4747994 0.53158115

1 -0.4316820 1.51412171

References

1. Comrey, A. L., & Lee, H. B. (2013). A first course in factor

analysis. Psychology Press.

2. Kim, J. O., & Mueller, C. W. (1978). Factor analysis: Statistical methods and practical issues (No. 14). Sage.
3. Choi, I., & Jeong, H. (2018). Model selection for factor analysis: Some new criteria and performance comparisons. Econometric Reviews, 1-20.
4. Chatfield, C. (2018). Introduction to multivariate analysis. Routledge.
5. Fan, J., Sun, Q., Zhou, W. X., & Zhu, Z. (2018). Principal component analysis for big data. arXiv preprint arXiv:1801.01602.
6. Crawley, M. J. (2012). The R book. John Wiley & Sons.
7. Anderberg, M. R. (1973). Cluster analysis for applications (No. OAS-TR-73-9). Office of the Assistant for Study Support Kirtland AFB N MEX.
8. Wilks, D. S. (2011). Cluster analysis. In International geophysics (Vol. 100, pp. 603-616). Academic press.
9. Romesburg, C. (2004). Cluster analysis for researchers. Lulu. com.
10. Sarstedt, M., & Mooi, E. (2014). Cluster analysis. In A concise guide to market research (pp. 273-324). Springer, Berlin, Heidelberg.
11. Morrison, D. G. (1969). On the interpretation of discriminant analysis. Journal of marketing research, 156-163.
12. Izenman, A. J. (2013). Linear discriminant analysis. In Modern multivariate statistical techniques (pp. 237-280). Springer, New York, NY.
13. Tharwat, A., Gaber, T., Ibrahim, A., & Hassanien, A. E. (2017). Linear discriminant analysis: A detailed tutorial. AI communications, 30(2), 169-190.
14. Cox, D. R. (2018). Analysis of survival data. Routledge.
15. Allignol, A., & Latouche, A. (2018). CRAN task view:

Survival analysis.

16. Miller Jr, R. G. (2011). Survival analysis (Vol. 66). John Wiley & Sons.
17. Klein, J. P., & Moeschberger, M. L. (2006). Survival analysis: techniques for censored and truncated data. Springer Science & Business Media.
18. Kleinbaum, D. G., & Klein, M. (2010). Survival analysis (Vol. 3). New York: Springer.
19. Collett, D. (2015). Modelling survival data in medical research. Chapman and Hall/CRC.
20. Crowther, M. J., & Lambert, P. C. (2014). A general framework for parametric survival analysis. Statistics in medicine, 33(30), 5280-5297.
21. Ishak, K. J., Kreif, N., Benedict, A., & Muszbek, N. (2013). Overview of parametric survival analysis for health-economic applications. Pharmacoeconomics, 31(8), 663-675.
22. Fox, J. (2002). Cox proportional-hazards regression for survival data. An R and S-PLUS companion to applied regression, 2002.

Also by Editor IJSMI

Software

R Programming - a Comprehensive Guide

Introduction To Business Statistics Through R Software

Standalone

Digital Technologies – an Overview of Concepts, Tools and Techniques Associated with it

Clinical Trial Management – an Overview

Bayesian Methodology: an Overview With The Help Of R Software

Forecasting Models – an Overview With The Help Of R Software

Introduction To Non Parametric Methods Through R Software

Watch for more at www.ijsmi.com/book.php.

About the Author

Editor,

International Journal of Statistics and Medical Informatics
www.ijsmi.com/book.php
editorijsmi@gmail.com
Read more at www.ijsmi.com/book.php.

www.ingramcontent.com/pod-product-compliance
Lightning Source LLC
LaVergne TN
LVHW090416210926
848480LV00002B/581